Study on Crack Resistance of Concrete

Study on Crack Resistance of Concrete

Guest Editor
Weiting Xu

Basel • Beijing • Wuhan • Barcelona • Belgrade • Novi Sad • Cluj • Manchester

Guest Editor
Weiting Xu
School of Materials Science
and Engineering
South China University of
Technology
Guangzhou
China

Editorial Office
MDPI AG
St. Alban-Anlage 66
4052 Basel, Switzerland

This is a reprint of the Special Issue, published open access by the journal *Materials* (ISSN 1996-1944), freely accessible at: www.mdpi.com/journal/materials/special_issues/C34VV7L1B0.

For citation purposes, cite each article independently as indicated on the article page online and as indicated below:

Lastname, A.A.; Lastname, B.B. Article Title. *Journal Name* **Year**, *Volume Number*, Page Range.

ISBN 978-3-7258-1192-2 (Hbk)
ISBN 978-3-7258-1191-5 (PDF)
https://doi.org/10.3390/books978-3-7258-1191-5

© 2024 by the authors. Articles in this book are Open Access and distributed under the Creative Commons Attribution (CC BY) license. The book as a whole is distributed by MDPI under the terms and conditions of the Creative Commons Attribution-NonCommercial-NoDerivs (CC BY-NC-ND) license (https://creativecommons.org/licenses/by-nc-nd/4.0/).

Contents

Preface . . . . **vii**

Xiaotong Xing, Weiting Xu, Guihua Zhang and Xilian Wen
The Mechanical Performance and Reaction Mechanism of Slag-Based Organic–Inorganic Composite Geopolymers
Reprinted from: *Materials* **2024**, *17*, 734, https://doi.org/10.3390/ma17030734 . . . . **1**

Qingyu Cao, Juncheng Zhou, Weiting Xu and Xiongzhou Yuan
Study on the Preparation and Properties of Vegetation Lightweight Porous Concrete
Reprinted from: *Materials* **2024**, *17*, 251, https://doi.org/10.3390/ma17010251 . . . . **20**

Feifei Jiang, Juan Zhou, Zhongyang Mao and Bi Chen
Study on the Influence of Magnesite Tailings on the Expansion and Mechanical Properties of Mortar
Reprinted from: *Materials* **2023**, *16*, 7082, https://doi.org/10.3390/ma16227082 . . . . **35**

Xuefeng Zhao, Zhongyang Mao, Xiaojun Huang, Penghui Luo, Min Deng and Mingshu Tang
Effect of Curing Conditions on the Hydration of MgO in Cement Paste Mixed with MgO Expansive Agent
Reprinted from: *Materials* **2023**, *16*, 4032, https://doi.org/10.3390/ma16114032 . . . . **51**

Zemeng Guo, Lingling Xu, Shijian Lu, Luchao Yan, Zhipeng Zhu and Yang Wang
Study on the Effect of PVAc and Styrene on the Properties and Microstructure of MMA-Based Repair Material for Concrete
Reprinted from: *Materials* **2023**, *16*, 3984, https://doi.org/10.3390/ma16113984 . . . . **63**

Xuan Zhou, Zhongyang Mao, Penghui Luo and Min Deng
Effect of Mineral Admixtures on the Mechanical and Shrinkage Performance of MgO Concrete
Reprinted from: *Materials* **2023**, *16*, 3448, https://doi.org/10.3390/ma16093448 . . . . **79**

Tian Liang, Penghui Luo, Zhongyang Mao, Xiaojun Huang, Min Deng and Mingshu Tang
Effect of Hydration Temperature Rise Inhibitor on the Temperature Rise of Concrete and Its Mechanism
Reprinted from: *Materials* **2023**, *16*, 2992, https://doi.org/10.3390/ma16082992 . . . . **94**

Zhe Cao, Zhongyang Mao, Jiale Gong, Xiaojun Huang and Min Deng
Effect of Working Temperature Conditions on the Autogenous Deformation of High-Performance Concrete Mixed with MgO Expansive Agent
Reprinted from: *Materials* **2023**, *16*, 3006, https://doi.org/10.3390/ma16083006 . . . . **106**

Jun Chen, Zhongyang Mao, Xiaojun Huang and Min Deng
Effect of the Water-Binder Ratio on the Autogenous Shrinkage of C50 Mass Concrete Mixed with MgO Expansion Agent
Reprinted from: *Materials* **2023**, *16*, 2478, https://doi.org/10.3390/ma16062478 . . . . **121**

Pengfei Li, Zhongyang Mao, Xiaojun Huang and Min Deng
Preparation and Performance of Repair Materials for Surface Defects in Pavement Concrete
Reprinted from: *Materials* **2023**, *16*, 2439, https://doi.org/10.3390/ma16062439 . . . . **134**

Anqun Lu, Wen Xu, Qianqian Wang, Rui Wang and Zhiyuan Ye
Effect of Different Expansive Agents on the Deformation Properties of Core Concrete in a Steel Tube with a Harsh Temperature History
Reprinted from: *Materials* **2023**, *16*, 1780, https://doi.org/10.3390/ma16051780 . . . . **153**

Preface

Concrete cracking is a crucial factor that threatens the durability and strength of concrete. Especially for concrete building structures with complex working environments such as hydraulic structures and long-span building structures, concrete cracking seriously threatens the safe operation of engineering projects. The cracking resistance of concrete is influenced by many factors, such as mechanical properties, temperature processes, autogenous shrinkage, restrained stress, creep, etc. The study of the crack resistance of concrete is of great significance for its wide application. This reprint focuses on the research on the cracking resistance of concrete, including the formation mechanism of concrete cracking, the relationship between the cracking mechanism and the performance of various types of concrete, and the resistance and maintenance of concrete cracks.

Weiting Xu
Guest Editor

Article

The Mechanical Performance and Reaction Mechanism of Slag-Based Organic–Inorganic Composite Geopolymers

Xiaotong Xing [1], Weiting Xu [1,*], Guihua Zhang [1] and Xilian Wen [2]

1. School of Materials Science and Engineering, South China University of Technology, Guangzhou 510641, China; 202010103718@mail.scut.edu.cn (X.X.); zhangguihuascut@163.com (G.Z.)
2. Guangzhou Residential Construction Development Co., Ltd., Guangzhou 510075, China; langer123456@126.com
* Correspondence: xuweiting@scut.edu.cn

Abstract: A series of organic–inorganic composite geopolymer paste samples were prepared with slag-based geopolymer and three types of hydrophilic organic polymers, i.e., PVA, PAA, and CPAM, by ordinary molding and pressure-mixing processes. The reaction mechanism between slag-based geopolymer and organic polymers was studied by FT-IR, NMR, and SEM techniques. The experimental results showed that the slag-based geopolymer with the addition of 3% PVA presented the highest 28-day flexural strength of 19.0 MPa by means of a pressure-mixing process and drying curing conditions (80 °C, 24 h) compared with the geopolymers incorporating PAA and CPAM. A more homogeneous dispersion morphology was also observed by BSE and SEM for the 3% PVA-incorporated slag-based geopolymer. The FT-IR testing results confirmed the formation of a C–O–Si (Al) bond between PVA and the slag-based geopolymer. The deconvolution of the Q^3 and $Q^2(1Al)$ species obtained by ^{29}Si NMR testing manifested the addition of PVA and increased the length of the silicon backbone chain in the geopolymer. These findings confirmed that a composite geopolymer with high toughness can be produced based on the interpenetrating network structure formed between organic polymers and inorganic geopolymer.

Keywords: mechanical performance; reaction mechanism; slag-based geopolymer; organic–inorganic composite geopolymer; pressure-mixing process

1. Introduction

Slag-based geopolymer is one of the most promising new green cementitious materials due to its advantages of low energy consumption, small expansion, and strong corrosion resistance [1–3]. The active silico–aluminate materials are mixed with a strong alkaline solution such as an alkali metal (Na, K) hydroxide or silicate to synthesize geopolymer. In this environment, a three-dimensional network structure of geopolymers is formed due to the rapid dissolution and reunion reaction of silica–aluminum active substances. The geopolymer structure mainly contains three different sialate units [4], namely, poly sialate (–Si–O–Al–), poly sialate-siloxo (–Si–O–Al–O–Si–), and polysialate-disiloxo (–Si–O–Al–O–Si–O–Si–), shown in Figure 1. The properties of the final products are highly dependent on the cross-linking degree of the different silico–aluminate polymeric chains [5].

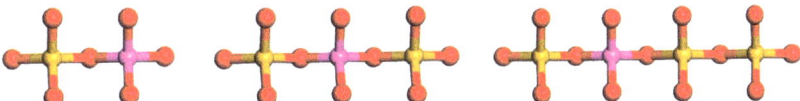

Figure 1. Structural unit model of geopolymer. (The red ball represents oxygen atom, the yellow ball represents the silica atom, and the purple ball represents the aluminum atom).

However, the low flexural strength and low fracture toughness of slag-based geopolymer limit its wider application in the construction field. It was found that when steel slag was used as the main aluminosilicate material to prepare geopolymer mortar, the flexural strength was only 11.7 Mpa while the compressive strength was 84 Mpa at 28 days [6]. In a study conducted by Kim et al. [7], the coefficient of flexural toughness was only 0.55 J. The defects of low toughness can be improved by making geopolymer composites. Many types of fillers such as various kinds of short and continuous fibers have been introduced into the geopolymer matrix to improve the mechanical properties [1,8,9]. Sükrü Özkan et al. [10] examined the effect of the hybrid use of 75% PVA fiber + 25% basalt fiber as additives on the mechanical performance of geopolymer concrete. The use of fiber in concrete significantly improved the resistance to the formation of cracks and contributed to increased ductility and the development of the mechanical strength and energy absorption of the concrete [11]. However, the reinforcement from a single species of fiber only physically improved the performance and crack resistance of the geopolymer matrix on one scale. The toughness of the geopolymer matrix itself was not improved. In addition, the compatibility of the fiber and geopolymer matrix also limited the construction and applications of the resulting composite materials.

It is noteworthy that a cement–polymer composite called "macro-defect-free (MDF) cement" with remarkably high toughness emerged in the early 1980s [12–14]. MDF cement is mainly composed of inorganic cement, water, and small amounts of organic polymer. The calcium aluminate cement–polyvinyl acetate polymer (CAC-PVAc) composite is one of the examples characterized by very high flexural strength of up to 70 MPa [15,16]. Based on micro-structural analysis, PVAc hydrolyzed under the high PH environment of the cement solution. Acetate ions dissolved from PVAc and subsequently reacted with the dissolved calcium ions from the calcium silicate in the cement to form calcium acetate [12,17] and, hence, formed an organic and inorganic bonding structure with high toughness.

The incorporation of various kinds of water-soluble polymers has been proven to improve the toughness and crack resistance of the cement and geopolymer matrix [18–22]. Catauroa [19] et al. introduced different proportions of polyethylene glycol (PEG) into metakaolin-based geopolymer and it was found that the PEG content and aging time affected the mechanical performance of geopolymers. The study also revealed that PEG leads to a network reorganization by increasing Al–O–Si bonds in the geopolymer matrix and forming H-bonds with the inorganic phase. Olesia Mikhailova [20] drew the conclusion that the maximum compressive and flexural strength of geopolymer is obtained with addition of 10% PEG400 because of the organic–inorganic bonding structure with high density and few pores. From the previous relevant studies, we have two points that need to be further studied and discussed.

First, the traditional common molding method may have a limited promotion effect on the bonding reaction of organic and inorganic compounds, and a better molding method needs to be explored. Second, the polymerization mechanism of water-soluble polymer and geopolymer needs more detailed and in-depth study.

The aim of this research is to investigate the influence of organic polymers with different functional groups (–OH, –COOH, and –$CONH_2$) on the performances of slag-based geopolymers and select a suitable organic polymer with a sound modification effect on the toughness of slag-based geopolymer. In this study, we incorporated three types of water-soluble polymers, namely, PVA, PAA, and CPAM, as organic additives into slag-based geopolymer to investigate the mechanical properties and polymerization reaction mechanism of the organic–inorganic hybrid composite geopolymer. To strengthen the organic–inorganic bonding reaction, the pressure-mixing process was adopted and compared to the traditional common modeling method in the process of making geopolymer paste specimens.

2. Materials and Methods

2.1. Materials

The slag used in this study was the S95 mine powder from Ma'anshan Tianrui slag trading Co., Ltd., Ma'anshan, China. The chemical composition of the slag examined by XRF is shown in Table 1. The particle size of the slag was analyzed by a laser particle size meter, and the particle size distribution is shown in Figure 2.

Table 1. Chemical compositions of slag.

Components (%)	CaO	SiO$_2$	Al$_2$O$_3$	MgO	SO$_3$	Fe$_2$O$_3$	TiO$_2$	MnO	LOI
slag	37.12	32.06	14.84	10.77	1.67	1.14	0.92	0.33	1.07

Note: LOI: Loss of ignition.

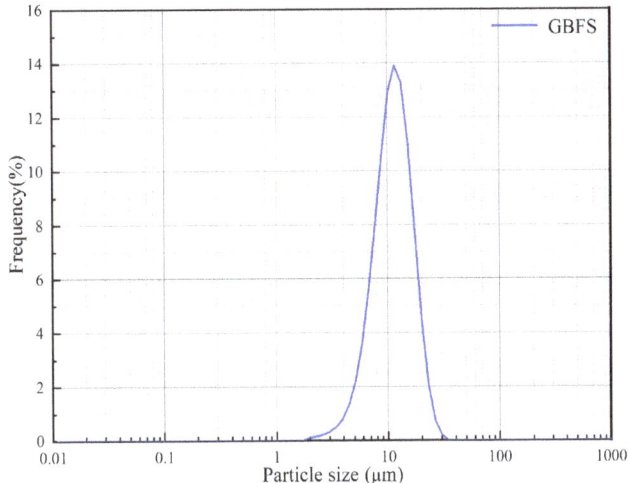

Figure 2. Particle size distributions of slag.

Based on the oxide components of slag in Table 1, the correlation quality indicators of the slag were calculated by the following Formulas (1)–(4) [23,24]. According to the calculation results, the slag was evaluated as being of a high-quality grade.

$$b = \frac{\text{CaO} + \text{MgO} + \text{Al}_2\text{O}_3}{\text{SiO}_2} = \frac{37.12 + 10.77 + 14.84}{32.06} = 1.96 > 1 \tag{1}$$

$$H_0 = \frac{\text{Al}_2\text{O}_3}{\text{SiO}_2} = \frac{14.84}{32.06} = 0.46 > 0.25 \tag{2}$$

$$M_0 = \frac{\text{CaO} + \text{MgO}}{\text{SiO}_2 + \text{Al}_2\text{O}_3} = \frac{37.12 + 10.77}{32.06 + 14.84} = 1.02 > 1 \tag{3}$$

$$K = \frac{\text{CaO} + \text{MgO} + \text{Al}_2\text{O}_3}{\text{TiO}_2 + \text{MnO} + \text{SiO}_2} = \frac{37.12 + 10.77 + 14.84}{32.06 + 0.33 + 0.92} = 1.88 > 1.2 \tag{4}$$

The alkali activator used in this study was water glass (industrial-grade sodium silicate) provided by Foshan Shengjing New Material Company (Foshan, China). The modulus of the water glass was 2.44 and the solid content was 46.17%, with 13.70% Na$_2$O and 32.47% SiO$_2$. Sodium hydroxide (analytical purity, Tianjin Jinhui Taiya Chemical Reagent Co., Ltd., Tianjin, China) was adopted to adjust the modulus of the alkaline activator.

The water-soluble organic polymers used in this study were cation polyacrylamides (CPAM), polyvinyl alcohol (PVA), and polyacrylic acid (PAA). They were purchased from

Shanghai Macklin Chemical Reagent Co., Ltd. (Shanghai, China). Both PVA and CPAM were granular. During the test, they were respectively pre-mixed with water to form a solution with 55% concentration for subsequent experiments. PAA was a liquid with a concentration of 50%.

2.2. Mixing Proportions

Before preparing organic and inorganic composite geopolymers, to achieve a suitable slag-based geopolymer with high mechanical performance as the inorganic substrate, the effects of the type, modulus, and concentration of alkali activators on the strength and polymerization degree of slag-based geopolymer are investigated in Section 3.1.

The mixing proportions of organic–inorganic composite geopolymer pastes are given in Table 2. It is worth pointing out that the proportions of water-soluble polymers added were 0.5%, 1%, and 2% for the common molding method and 1%, 3%, and 5% for the pressure-mixing process.

Table 2. Mixture proportions of pastes in weight (%).

Mix ID	Polymer Type		Mix Proportions			
	Species	Weight	Water	Slag	$BaCl_2$	$(Na_2PO_3)_6$
Reference	-	-	0.35	1	0.01	0.005
A1	PVA	0.005	0.35	1	0.01	0.005
A2	PVA	0.01	0.35	1	0.01	0.005
A3	PVA	0.02	0.35	1	0.01	0.005
B1	PAA-Na	0.005	0.35	1	0.01	0.005
B2	PAA-Na	0.01	0.35	1	0.01	0.005
B3	PAA-Na	0.02	0.35	1	0.01	0.005
C1	CPAM	0.005	0.35	1	0.01	0.005
C2	CPAM	0.01	0.35	1	0.01	0.005
C3	CPAM	0.02	0.35	1	0.01	0.005

2.3. Ordinary Molding Process

The water-to-solid ratio of the geopolymer pastes was 0.35. All geopolymer pastes were mixed in a Hobart mixer at an ambient temperature. The alkaline solution was prepared by an alkali activator and water in advance. Then, the slag was added to the alkaline solution and stirred for 4 min. Finally, the water-soluble polymer was added into the geopolymer slurry and slowly stirred for 6 min. All pastes were cast with sizes of 25 mm × 25 mm × 140 mm and 20 mm × 20 mm × 20 mm for flexural strength and compressive strength tests, respectively. For the first 24 h, the pastes were sealed with a layer of polyethylene to prevent the evaporation of moisture. Then, they were demolded, tightly sealed in plastic bags, and cured in a steam-curing chamber with a temperature of 20 ± 1 °C and 99% relative humidity until the testing dates.

2.4. Pressure-Mixing Process

To promote the breaking and repolymerization of more chemical bonds in organic molecules and aluminosilicates in geopolymers, the pressure-mixing process was used. The mixer parameter was set to the front and rear roller roll ratio of 1:3, and the roll distance was 1.5 mm. The plastic organic–inorganic composite geopolymer slurry was subjected to mixing for 5 min. The obtained mixture was combined by passing it repeatedly through the nip of two steel roller mills at narrow nip gaps and, subsequently, a higher shearing was applied to the mixture with higher rotating speeds. A paste sheet was laminated in a mold with a size of 120 mm × 30 mm × 10 mm and molded by a universal testing machine. In addition, to explore the influence of curing conditions on strength, the pastes were sealed by a layer of polypropylene after being demolded for 24 h. Then, half of the pastes were placed in a steam-curing chamber with 99% relative humidity at 80 °C for 24 h, whereas the remaining pastes were placed in an oven and cured at 80 °C for 24 h.

2.5. Mechanical Properties of the Pastes

Pastes were cast into 25 mm × 25 mm × 140 mm, 10 mm × 30 mm × 120 mm, and 20 mm × 20 mm × 20 mm molds for mechanical property tests [25]. The flexural and compressive strength of the pastes was measured by a fully automatic universal testing machine at the ages of 3 d, 7 d, and 28 d. Flexural strength tests were carried out using the three-point bending method at a loading rate of 2 mm/min. The strength result was based on the average of three testing pastes for each mix.

2.6. The Characterization of Organic–Inorganic Composite Material

2.6.1. Fourier Transform Infrared Reflection (FTIR) Tests

FT-IR (Vertex 70 produced by Bruker, Ettlingen, Germany) tests were conducted in a transmission model with a scan range of 4000–400 cm^{-1}. The KBr tablet method was used in the test. The ratio of the powder sample to dry KBr was taken from 1:100 to 1:200 in an agate mortar for 5 min, and the powder was put into a grinding mixture after no reflection. The piezoelectric sheet was pressed internally and, finally, the transparent circular sheet was tested by infrared spectra.

2.6.2. Solid-State ^{29}Si Nuclear Magnetic Resonance (NMR) Tests

Avance III PULS 400 MHz NMR equipment, produced by Bruker, Germany, was used to identify and quantify different types of silicon. The resonance frequency was 79.3 mHz, the rotation angle was $\pi/2$, the pulse width was 4 µs, and the magic angle rotation frequency was 8 kHz. In addition, PeakFit 4.0 software was used for the peak fitting of the original map for data processing in this experiment. To accurately separate peak results, the AutoFit Peaks III Deconvolution program was used, and the relevant parameter r2 was greater than 0.99.

2.6.3. Inductively Coupled Plasma Optical Emission Spectrometry (ICP-OES) Tests

The concentrations of Si, Ca, and Al ions in the solution were determined by the ICP-OES method. The mixing ratio of the test solution is shown in Table 3. At 30 min, 60 min, 90 min, 120 min, and 180 min, with centrifugation at 3000 rpm for 10 min, the solution was separated and collected from the paste, and the clean solution was obtained after filtration. The plasma gas flow rate was 15 L/min. The atomized gas flow rate was 0.6 L/min. The pump speed was 100 rpm.

Table 3. Mixture proportions for preparing samples in large water–cement ratios.

Alkali Content (%)	NaOH (g)	Slag (g)	Water (g)
2	5.36	20	70
4	10.84	20	70
6	16.12	20	70
8	21.50	20	70

2.6.4. Backscattered Electron (BSE) and Scanning Electron Microscope (SEM) Tests

The morphological characteristics and element distribution of the geopolymer pastes were assessed using SEM (SU8220, HITACHI, Tokyo, Japan) with BSE. The working distance was 12 mm and the voltage was 10 kV under the secondary electron (SE) mode.

2.6.5. Mercury Intrusion Porosimetry (MIP) Tests

The pore size distribution of the pastes was determined using the MIP method with the Micromeritics Auto Pore IV 9500 mercury injection instrument (Micromeritics Instrument Corporation, Norcross, GA, USA). The test pressure range of the MIP was 0.0036–210 MPa.

3. Results and Discussion

3.1. The Effects of Different Alkali Contents and Moduli on the Properties of Slag-Based Geopolymer

3.1.1. Mechanical Properties of Pastes

The compressive strength and the flexural strength of slag-based geopolymer pastes with different alkali activator (sodium hydroxide) contents are shown in Figure 3. The highest compressive and flexural strength was observed for pastes with 6% alkali content.

For the geopolymer with water glass as the alkali activator, the effect of the modulus of water glass on the compressive and flexural strength of the pastes is shown in Figure 4. It can be observed that pastes with a modulus of 1.6 and 6% alkali content presented the highest compressive strength of 91.0 Mpa and flexural strength of 7.5 Mpa at 28 d, respectively. Comparing the results of Figures 3 and 4, it is more appropriate to choose water glass as the activator for geopolymer pastes.

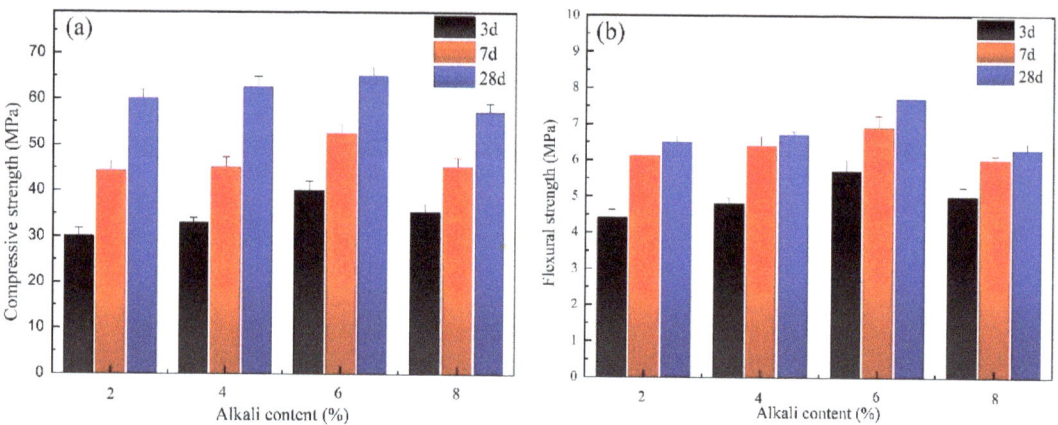

Figure 3. Effect of alkali content on mechanical properties of slag-based geopolymer paste: (**a**) compressive strength; (**b**) flexural strength.

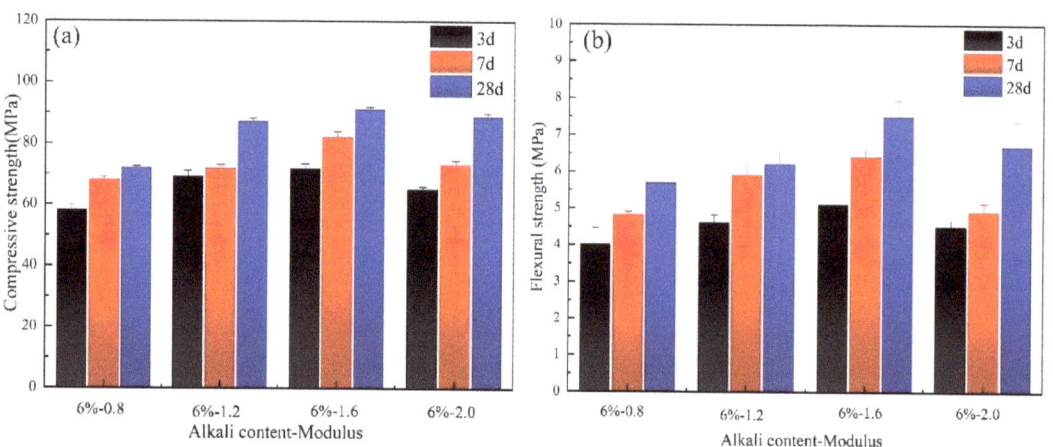

Figure 4. Effect of modulus on mechanical properties of slag-based geopolymer: (**a**) compressive strength; (**b**) flexural strength.

3.1.2. ICP Analysis

The evolution of Si^{4+}, Ca^{2+}, and Al^{3+} concentrations in the solutions of slag pastes with different alkali contents is shown in Figure 5. As the alkali content increased, the high dissolution concentration of Si^{4+} reduced over time. This was caused by more OH^- and Na^+ in the solution, which were available in the network structure of the vitreous body. Then, Si^{4+} gradually formed $[SiO_4]^{4-}$ combined with $[AlO_4]^{5-}$ and Ca^{2+}. Ca^{2+} and Al^{3+} concentrations with a 6% alkali content were extremely high, which was beneficial for the polymerization reaction. The dissolution–polymerization process equilibrated to form more hydration products. The concentration changes of Si^{4+}, Ca^{2+}, and Al^{3+} in the slag paste solutions with different alkali contents over time are shown in Figure 5. The dissolved concentration of Si^{4+} decreased with the extension of time and increased with the increase in alkali content. This was due to the fact that more OH^-, Ca^{2+}, and Al^{3+} dissolved in the solution were available in the reticular structure that formed the vitreous body. Si^{4+} then gradually formed $[SiO_4]^{4-}$ and $[AlO_4]^{5-}$, which bonded with Ca^{2+}. When the alkali content was 6%, the concentrations of Ca^{2+} and Al^{3+} in the solution reached the highest values, which was conducive to promoting the polymerization reaction and forming more polymerization products.

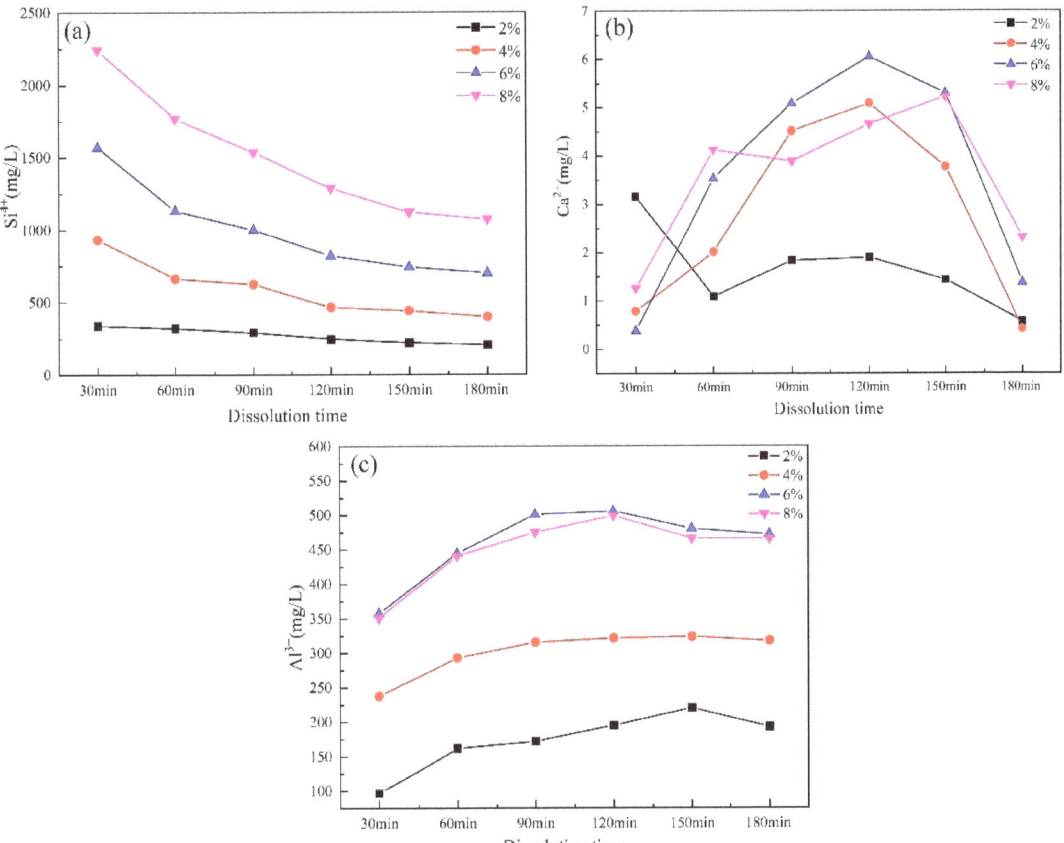

Figure 5. Ion concentration of slag in slurry with large water–slag ratios and different alkali contents: (**a**) Si^{4+}; (**b**) Ca^{2+}; (**c**) Al^{3+}.

3.1.3. FTIR Analysis

To study the effect of alkali concentration on the chemical bond of slag-based geopolymer, paste samples were subjected to FTIR analysis. The results are shown in Figure 6. At 1003~1035 cm^{-1}, there was a large diffuse band that lacked periodicity, representing the characteristic of amorphous Si-Al polytetrahedral. For the geopolymers with 6% alkali content, the Si-O-Si peaks at 460 cm^{-1} and the Si-O-Al peaks at 960 cm^{-1} were sharper and more intense. This can be attributed to more polycondensation occurring between [SiO$_4$]$^{4-}$ and [AlO$_4$]$^{5-}$ [26]. In addition, geopolymers with 6% alkali content were found to have more strength at 891 cm^{-1}, 1021 cm^{-1}, and 1097 cm^{-1}, representing the diverse characteristic groups of C-(A)-S-H polymerization products (SiO$_4$)$^{4-}$, (SiO$_3$)$^{2-}$, and SiO$_2$, respectively. It is apparent that an alkali content of 6% is the most favorable alkaline environment to promote the polymerization of slag-based geopolymer [27].

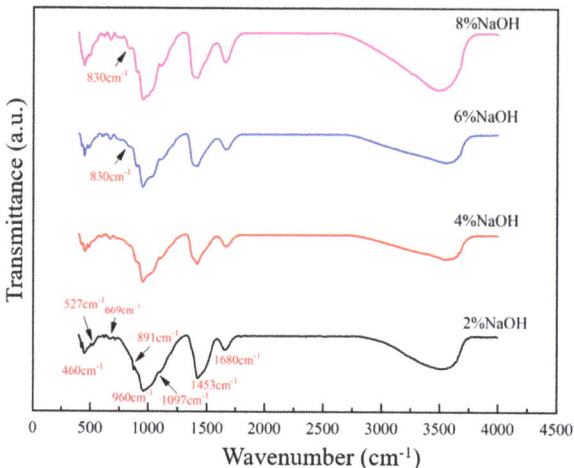

Figure 6. FTIR spectra of slag-based geopolymers prepared with NaOH activators at 28 d with different alkali contents.

3.1.4. ^{29}Si NMR Analysis

The polymeric structure of geopolymer was studied by the ^{29}Si nuclear magnetic resonance technique. The distribution of the polymerization degree and main chain length of slag-based polymers prepared with different alkali contents and different moduli of activators are shown in Tables 4 and 5. The polymerization degree analysis showed that the cumulative strength of Q^0 was the smallest when the alkali content was 6%, which indicated that more slag was consumed. The average chain length of C-(A)-S-H in the hydration product could be calculated by Equation (5) [28]. When the base content was 6% and the modulus was 1.6, the average chain length was the largest.

$$MCL = \frac{I(Q^1(0Al)) + I(Q^2(0Al)) + \frac{3}{2}I(Q^2(1Al))}{\frac{1}{2}I(Q^1(0Al))} \quad (5)$$

Table 4. Polymerization distribution and main chain length of slag-based geopolymers prepared with NaOH activators with different alkali contents (28 d): (a) 2% NaOH; (b) 4% NaOH; (c) 6% NaOH; (d) 8% NaOH.

Sample	Q^n Cumulative Strength (I/%)							MCL
	Q^0	Q^1(1Al)	Q^1(0Al)	Q^2(1Al)	Q^2(0Al)	Q^3(1Al)	Q^3(0Al)	
(a)	18.91	22.73	20.01	24.23	10.37	-	3.75	6.12
(b)	16.96	23.25	20.59	25.05	9.09	3.67	1.39	6.51
(c)	14.06	25.2	21.62	27.52	9.44	2.10	0.09	6.69
(d)	15.30	26.56	24.19	26.86	7.09	-	-	5.91

Table 5. Degree of polymerization distribution and main chain length of slag-based geopolymers prepared with different modulus activators (28 d): (a) 6%−0.8; (b) 6%−1.2; (c) 6%−1.6; (d) 6%−2.0.

Sample	Q^n Cumulative Strength (l/%)							MCL
	Q^0	$Q^1(1Al)$	$Q^1(0Al)$	$Q^2(1Al)$	$Q^2(0Al)$	$Q^3(1Al)$	$Q^3(0Al)$	
(a)	16.45	15.44	21.33	29.53	11.60	3.68	1.97	6.81
(b)	13.67	17.31	18.85	26.54	15.22	5.15	2.96	7.83
(c)	7.64	19.81	17.59	24.44	19.33	6.18	5.01	8.36
(d)	6.53	21.89	25.24	-	36.32	-	10.02	7.49

3.2. Effect of Water-Soluble Polymer Type on Slag-Based Geopolymer Paste

3.2.1. Mechanical Properties

The compressive strength and the flexural strength of slag-based geopolymer with three water-soluble polymers at the age of 3 d, 7 d, and 28 d are shown in Table 6. For all the mixtures, 6% water glass with a modulus of 1.6 was used as the alkali activator. PVA-0.5 showed the highest compressive strength and flexural strength among all the pastes. Under the same curing conditions, the PAA-Na and CPAM pastes presented lower flexural strength results than those of the reference mortar.

Table 6. Effect of polymer type and dosage on compressive and flexural strengths of paste.

	Compressive Strength (MPa)			Flexural Strength (MPa)		
	3 d	7 d	28 d	3 d	7 d	28 d
Reference	71.8	82	91	5.1	6.4	7.5
A1	77.7	89.6	98.3	6.9	8.3	9.2
A2	73.1	83.4	90.3	6	7.2	8.3
A3	60	70.7	76.2	5.3	6.1	7.3
B1	65	70.2	78.3	4.6	5.2	6.2
B2	62.4	64.3	71.5	4.4	4.9	5.8
B3	58.9	60	62.5	3.7	4.5	5.3
C1	63	65.6	70.9	4.8	5.9	6.6
C2	57.1	59.7	63.2	4.1	5.2	6
C3	51.7	57.8	60.1	4.4	4.7	5.5

3.2.2. FTIR Analysis

FTIR spectra of slag-based geopolymer composite pastes without and with the incorporation of 0.5%, 1%, and 2% PVA at 28 d are shown in Figure 7. The absorption peaks at 452 cm^{-1} representing the Si-O bending vibration bond gradually moved to the higher wave number with increases in the PVA dosage, indicated by the paste with 0.5% PVA (460 cm^{-1}), the paste with 1% PVA (475 cm^{-1}), and the paste with 2% PVA (489 cm^{-1}). The results show that the addition of PVA led to an increase in the Si–O tetrahedral polymerization structure in C–S–H [29–31].

The stretching modes of the SiO_n polymers and monomers with n = 4, 3, 2, 1, and 0 in geopolymer corresponded to the absorption bands at 1200 cm^{-1}, 1100 cm^{-1}, 950 cm^{-1}, 900 cm^{-1}, and 850 cm^{-1}, respectively. The silicon of SiO_n is prone to being replaced by aluminum due to the weak Al–O bond. With n from 4 to 0, the expansion pattern of the absorption band at 1200 cm^{-1}, 1100 cm^{-1}, 950 cm^{-1}, 900 cm^{-1}, and 850 cm^{-1} shifted to a lower wave number. It is worth pointing out that the absorption peak at 986 cm^{-1} was caused by the asymmetric tensile vibration of the Si–O–Si (Al) bond belonging to the characteristic C–Write in formal instead of italics Write in formal instead of italics

(A)–S–H [32]. With the increase in the PVA dosage, the absorption peak gradually moved towards the lower wave number, indicating that the addition of PVA promoted the binding of $[SiO_4]^{4-}$ and $[AlO_4]^{5-}$.

The vibration peak of –OH at 2933 cm^{-1} in PVA was overlapped by Si–OH in the polymerizate. The absorption peak at 3463 cm^{-1} was attributed to the asymmetric stretching vibration of the hydroxyl group [33], which broadened with the increase in PVA addition.

It is worth noting that when the PVA dosage was 2.0%, there was a weak absorption peak at 1180 cm^{-1} attributed to the C–O–Si (Al) vibration [34] formed by the organic–inorganic chemical reaction.

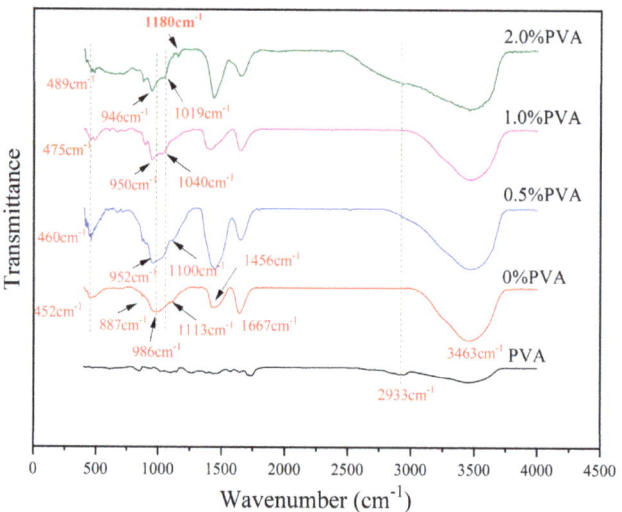

Figure 7. FTIR spectra of paste at 28 d.

3.2.3. ^{29}Si NMR Analysis

The ^{29}Si NMR spectra of the reference slag-based geopolymer paste and the geopolymer paste with 0.5% PVA are shown in Figure 8. The polymerization distribution, main chain length, and reaction degree of polymerization of the $[SiO_4]^{4-}$ tetrahedra in the paste at 28 d were calculated based on the results in Figure 8 and are shown in Table 7.

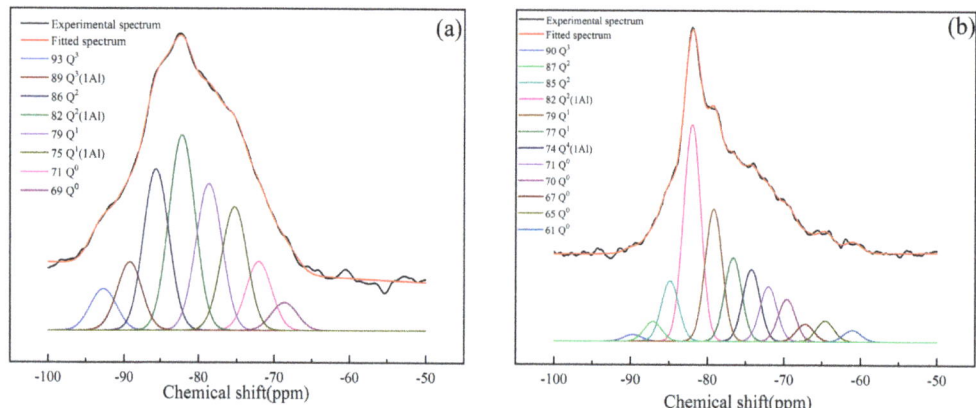

Figure 8. ^{29}Si NMR spectra of (**a**) reference geopolymer paste and (**b**) geopolymer paste with 0.5% PVA.

Table 7. Polymerization distribution, main chain length, and reaction degree of paste at 28 d.

PVA Content	Q^n Cumulative Strength (I/%)							MCL	α
	Q^0	Q^1(1Al)	Q^1(0Al)	Q^2(1Al)	Q^2(0Al)	Q^3(1Al)	Q^3(0Al)		
0%	7.64	19.81	17.59	24.44	19.33	6.18	5.01	8.36	82.85%
0.5%	13.98	17.77	19.22	29.31	18.85	-	0.87	8.70	73.14%

Compared with the reference paste, the paste with 0.5% PVA content did not detect the Q^3(1Al) silica tetrahedron, which may have been due to PVA preventing Al atoms from replacing Si atoms in the Q^3 silica tetrahedron or PVA grafting at a missing bridge silicon-tetrahedron site to form C–O–Si bonds. From the MCL results, the average chain length with 0.5% PVA was longer, and the Q^1 (0Al) and Q^2 (1Al) silicone tetrahedrons had a higher accumulation intensity, which indicated the presence of more chain structures in this paste. However, the accumulation intensity of the Q^3 (0Al) silicone tetrahedron was only 0.87%, which indicates that it did not easily form a high polymerization product with the addition of the PVA. However, according to reference [35], it is known that the main polymerization structure of influential properties is Q^1, Q^2, so the 0.5% PVA sample had better mechanical properties.

The reaction degree (α) of the geopolymer was estimated according to Equation (6). $I[Q^0]$ is proportional to the number of unreacted phases and $I[Q^1]$ and $I[Q^2]$ are proportional to the number of reaction products. From the calculation results shown in Table 7, the reaction degree reactivity of the reference paste was 82.85%, while that of 0.5% PVA was 73.14%. It is possible that the introduction of PVA reduced the degree of polymerization of aluminosilicate in the slag.

$$\alpha = \left[1 - \frac{I(Q^0)}{\sum_{i=0}^{2} I(Q^2)}\right] * 100\% \tag{6}$$

3.2.4. SEM Analysis

The morphologies of the reference geopolymer paste, geopolymer paste with 0.5% PVA, and geopolymer paste with 2.0% PVA at 28 d are shown in Figure 9. It apparent that the hydration product of the reference geopolymer paste was relatively large. The hydration products of geopolymer paste containing 0.5% PVA exhibited a scale-like structure, and the bonds between scales were very tight to form a whole structure. When the amount of PVA in the geopolymer was increased to 2.0%, the hydration products formed thin sheets and the geopolymer matrix turned fluffy. It is apparent that the excessive addition of PAV may lead to a decrease in the homogeneity and mechanical strength of the matrix.

Figure 9. Morphology of (a) reference geopolymer paste, (b) geopolymer paste with 0.5% PVA, and (c) geopolymer paste with 2.0% PVA at 28 d.

3.3. The Effect of the Pressure-Mixing Process of Pastes with Various Dosages of PVA under Different Curing Conditions

3.3.1. The Mechanical Properties of Pastes with Various Dosages of PVA Prepared by the Pressure-Mixing Process

The flexural strength of geopolymer pastes with 1%, 3%, and 5% PVA content obtained by the pressure-mixing process is shown in Figure 10. The incorporation of 3% PVA noticeably increased the flexural strength of pastes, with maximum values of 10.9 MPa at 3 d, 11.3 MPa at 7 d, and 12.6 MPa at 28 d, respectively. However, a further increase in the PVA dosage to 5.0% did not further improve the strength. An excessive increase in the amount of PVA led to the local agglomeration of organic polymer and reduced the effective site of the organic–inorganic polymerization reaction, which led to the weakening of organic–inorganic composite geopolymer. Thus, the dosage of 3% PVA was used for subsequent experiments. In contrast, using ordinary processes under the same curing conditions, the flexural strength began to decrease when the PVA content was increased by 1% or more. These results indicate that the pressure-mixing process significantly improves the flexural strength.

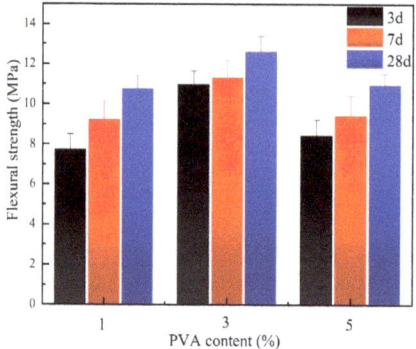

Figure 10. Flexural strength of pastes prepared by pressure-mixing process.

3.3.2. Mechanical Properties of Pastes with 3% PVA under Different Curing Conditions

The flexural strength of pastes with 3% PVA under different curing conditions are shown in Figure 11. Compared with the pastes cured under normal conditions (20 °C, 24 h), the flexural strengths of pastes under dry curing conditions (80 °C oven maintenance, 24 h) and steaming curing conditions (80 °C teaming maintenance, 24 h) were generally enhanced, especially for the ones under 80 °C dry conditions. Thus, dry curing is recommended for geopolymer pastes containing water-soluble polymers.

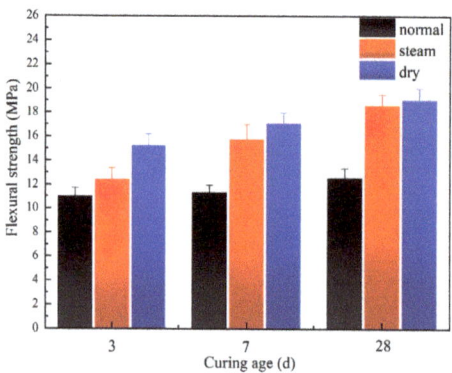

Figure 11. Flexural strength of pastes prepared by different curing conditions.

3.3.3. FTIR Analysis

The FTIR spectra of geopolymer pastes with different PVA dosages prepared by the pressure-mixing process with different curing regimes at 28 d are shown in Figure 12. As seen in Figure 12a, with an increase in the addition of PVA, the absorption peak of pastes at 473 cm^{-1} shifted slightly to the left, indicating that the degree of polymerization between the Si–O bonds decreased. The tensile vibration peak of the –OH bond was mainly at 3522 cm^{-1}, and the smaller absorption peak at 1655 cm^{-1} was the bending vibration peak of H–O–H. With an increase in PVA content, the peak position of both sites shifted to the right because the hydrogen bond formed by the binding water of the hydration product was reduced. Compared with the normal molding process, the characteristic peak of C–A–S–H near 1006 cm^{-1} was removed by about 50 wave numbers, indicating the formation of a high polymerization structure. According to Figure 12b, the main absorption peaks of pastes under different curing regimes in the spectra were basically consistent. However, the wave numbers of the main absorption peaks were obtained at 3498 cm^{-1}, and the peak shape was widened for pastes under steaming curing conditions, indicating an increase in binding water content and hydrogen bonds.

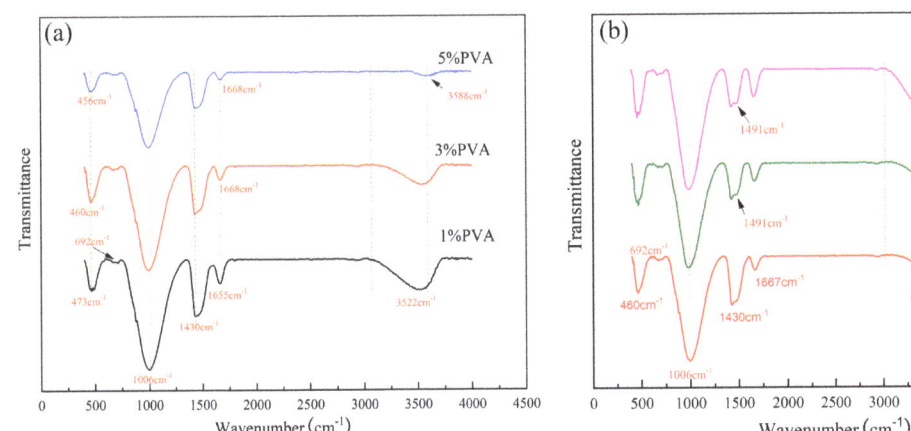

Figure 12. FTIR spectra of pastes prepared by pressure-mixing process with (**a**) different contents of PVA and (**b**) different curing regimes at 28 d.

3.3.4. ^{29}Si NMR Analysis of Pastes with Different Dosages of PVA Prepared by the Pressure-Mixing Process under Different Curing Regimes

The ^{29}Si NMR spectra of pastes prepared by the pressure-mixing process with 1% PVA under normal curing conditions, 3% PVA under normal curing conditions, 5% PVA under normal curing conditions, 3% PVA under dry heat curing conditions, and 3% PVA under steam curing conditions at 28 d are shown in Figure 13a–e, respectively. The polymerization distribution of pastes prepared by the pressure-mixing process under different curing regimes at 28 d are given in Table 8. Compared with Figure 13b,d,e, when the PVA content was 3%, the original spectrum of the high-temperature cured paste was sharper and less smooth. The position of the strongest peak gradually shifted to the lower direction of the magnetic field. Combined with the results in Table 8, it is apparent that Si in the Q^2 silico-oxygen tetrahedron in the paste was more easily replaced by a certain amount of Al atoms. The cumulative strength of Q^3 silica tetrahedrons cured at high temperatures was twice that at normal temperatures, indicating that high-temperature conditions promote the polycondensation reaction between silica tetrahedrons, which is conducive to the formation of a one-layer or double-chain high-polymerization structure.

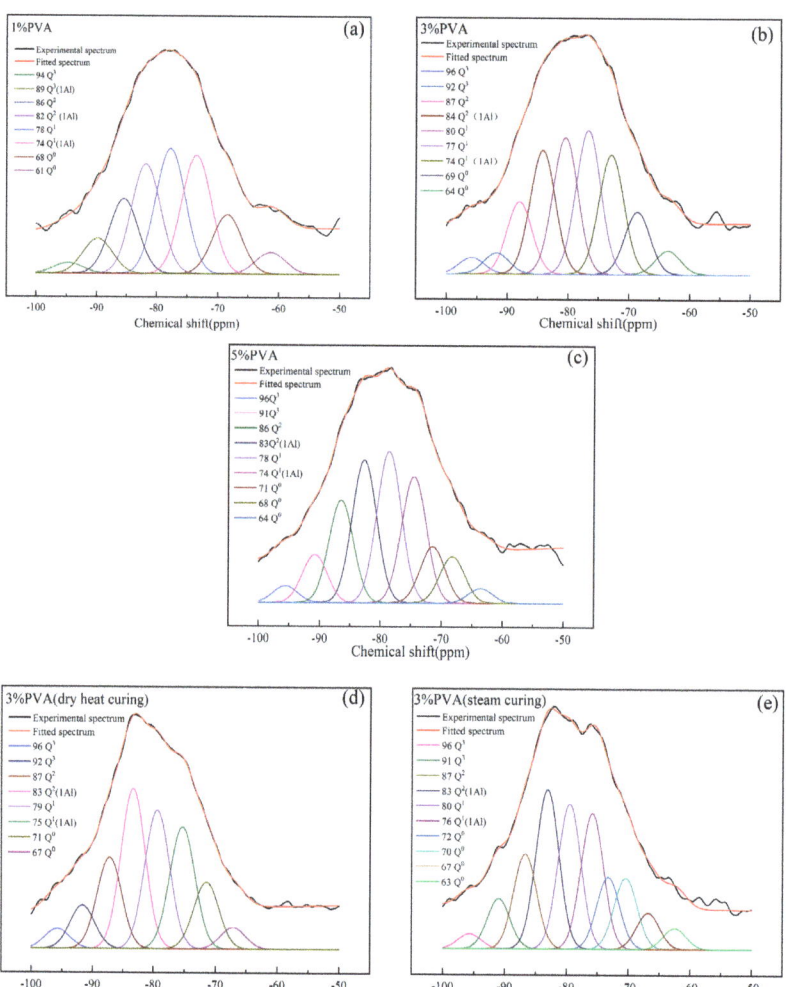

Figure 13. ^{29}Si NMR spectra of pastes prepared by pressure-mixing process with (**a**) 1% PVA (normal curing), (**b**) 3% PVA (normal curing), (**c**) 5% PVA (normal curing), (**d**) 3% PVA (dry heat curing), and (**e**) 3% PVA (steam curing) at 28 d.

Table 8. Polymerization distribution of pastes prepared by pressure-mixing process under different curing regimes at 28 d.

Content Curing Condition	Q^n Cumulative Strength (I/%)						
	Q^0	Q^1(1Al)	Q^1(0Al)	Q^2(1Al)	Q^2(0Al)	Q^3(1Al)	Q^3(0Al)
1%	14.59	21.39	24.57	17.79	17.48	2.31	1.87
3%	19.99	16.63	22.05	18.98	27.28	-	5.07
5%	21.73	18.88	19.45	20.20	15.35	-	4.39
3% dry	9.34	18.39	20.85	25.14	15.80	-	10.03
3% steam	10.07	8.99	16.91	19.50	33.72	-	10.51

3.3.5. Morphology Analysis
BSE Analysis

The BSE images of pastes prepared by the pressure mixing process with 1% PVA under normal curing conditions, 3% PVA under normal curing conditions, 5% PVA under normal curing conditions, 3% PVA under dry heat curing conditions, and 3% PVA under steam curing conditions at 28 d are shown in Figure 14a–e, respectively. Compared with Figure 14a–c, it can be observed that when the PVA dosage was 1%, no organic phase was found, while an organic phase was observed in the paste with a 5% PVA dosage, as shown in Figure 14c. For the pastes under normal curing conditions, with an increase in PVA content, cracks in the geopolymer matrix became obvious, maybe due to the different hardening rates of organic particles and slag-based geopolymer. Compared with Figure 14b,d,e, the paste under high-temperature curing had a relatively dense structure, and there was no obvious crack around the organic phase, but the dispersion was still uneven. In addition, by comparing Figure 14d,e, it can be observed that different high-temperature curing regimes had different effects on the PVA state in the matrix. With dry heat curing, PVA bonded more with slag geopolymer, and there was no obvious two-phase interface. For the paste under steaming curing conditions, the interface between the embedded PVA and the slag geopolymer matrix was obvious.

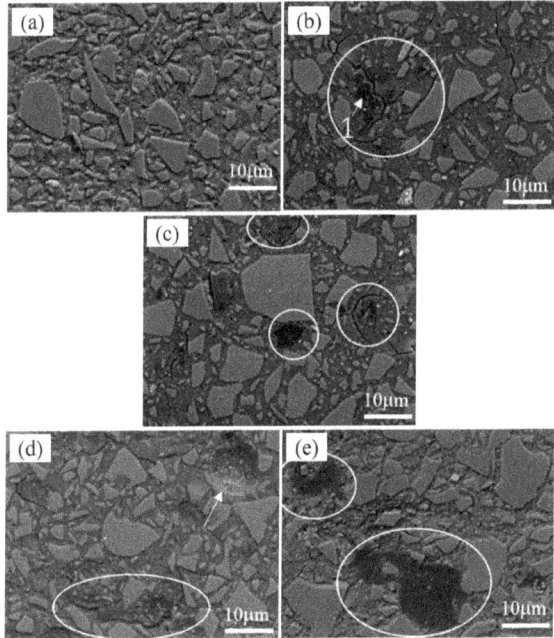

Figure 14. BSE images of pastes prepared by pressure mixing process with (**a**) 1% PVA (normal curing), (**b**) 3% PVA (normal curing), (**c**) 5% PVA (normal curing), (**d**) 3% PVA (dry heat curing), and (**e**) 3% PVA (steam curing) at 28 d.

SEM Analysis

The morphology images of pastes prepared by the pressure mixing process with 1% PVA under normal curing conditions, 3% PVA under normal curing conditions, 5% PVA under normal curing conditions, 3% PVA under dry heat curing conditions, and 3% PVA under steam curing conditions at 28 d are shown in Figure 15a–e, respectively. As shown in Figure 15a–c, with an increase in PVA dosage, no characteristic reaction product appeared, and the geopolymer matrix had a uniform texture, indicating that a good bond formed between the organic and inorganic phases under the pressure-mixing process. Compared with Figure 15b,d,e, the reaction product under different curing conditions had a different morphology. For the paste under dry curing conditions, there was a characteristic morphology of the internal structure of the pomegranate. This was because the PVA, which had lost water, wrapped some slag particles and rapidly shrunk under dry curing conditions. This kind of organic–inorganic composite structure is closely combined with the hydration product so that the material is not prone to brittle cracks during the loading process. For the paste under steam curing conditions, a variety of reaction products were produced in the dense matrix due to the early water retention effect.

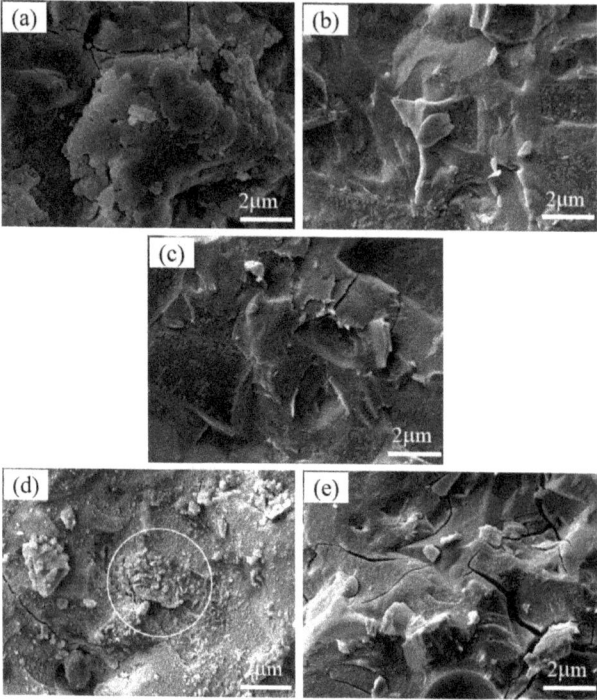

Figure 15. Morphology of pastes prepared by pressure mixing process with (**a**) 1% PVA (normal curing), (**b**) 3% PVA (normal curing), (**c**) 5% PVA (normal curing), (**d**) 3% PVA (dry heat curing), and (**e**) 3% PVA (steam curing) at 28 d.

3.3.6. Porosity and Pore Size Distribution of Pastes

The pore size distribution and cumulative pore volume of hardened pastes with 3% PVA prepared by different molding processes and curing regimes at 28 d are shown in Figure 16. Figure 16a shows that the order of the maximum pore size of each sample is paste prepared by ordinary molding (60.77 μm) > paste prepared by mixing molding > paste prepared by pressure mixing molding. The results show that the pressure mixing process eliminated some large pores in the paste matrix. However, the cumulative pore volume of each paste was contrary to the expected results, which may indicate that some pore defects were introduced during the pressure mixing process itself, resulting in an increase in the total pore volume of the paste.

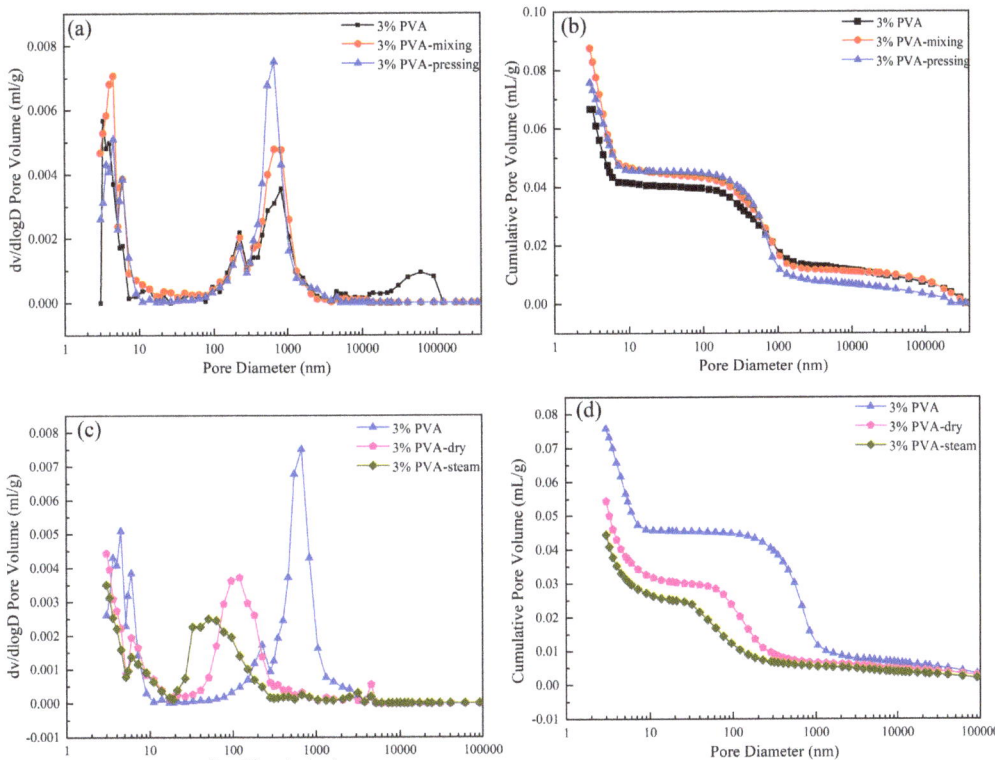

Figure 16. (**a**) Pore size distribution of pastes prepared by different molding processes at 28 d, (**b**) cumulative pore volume of pastes prepared by different molding processes at 28 d, (**c**) pore size distribution of pastes prepared by different curing regimes at 28 d, (**d**) cumulative pore volume of pastes prepared by different curing regimes at 28 d.

It is seen from Figure 16 that the order of maximum pore size and cumulative pore volume of each paste is as follows: paste under room temperature curing sample > paste under dry heat curing sample > paste under steam curing sample. The results show that under the high-temperature curing conditions, the reactivity of slag geopolymer was enhanced. More gel products were produced in the system, and the voids were gradually filled by slag particles, which reduced the pore size and made the matrix compact.

The maximum pore size and total pore volume of the pastes treated with steam curing were 50.34 nm and 0.0443 mL/g, respectively, which were the smallest among all the pastes. The reason for this is that steam curing continued to provide water for the unreacted slag particles in the later stage of the geopolymer reaction, resulting in more reaction products to fill the pores, refine the pores, and reduce the total pore volume of the paste. These results explain the reason for the maximum bending strength of the paste under steam curing conditions.

4. Conclusions

The effect of water-soluble polymers as additives on the mechanical and structural properties of slag-based geopolymer was investigated in this study. To reveal the modification mechanism of organic additives on slag-based geopolymer, the pore diameter distribution, boning, and polymerization degree of organic–inorganic composite geopolymer were explored. Based on this study, the following conclusions can be drawn:

1. For geopolymer paste using sodium silicate as an alkali activator (alkali content of 6% and modulus of 1.6) prepared by the ordinary production process under the same curing age, the flexural strength of the geopolymer paste with 0.5 wt% PVA was 9.2 Mpa, which was 29.33% higher than that of the reference geopolymer paste.
2. By means of the pressure-mixing molding process, the fraction of large capillary pores in the geopolymer pastes was significantly reduced. The flexural strength of the paste with 3% PVA was 12.6 Mpa at 28 d, while the flexural strength value increased markedly to 19.0 MPa due to the acceleration of the polymerization reaction by dry curing (80 °C, 24 h).
3. The FTIR spectrum of 2% PVA modified slag-based geopolymer shows that the vibration peak near 1180 cm^{-1} was due to the presence of a C–O–Si bond. The results show that the -OH functional group in C–(A)–S–H reacted with the –OH in PVA, which provided the possibility for the formation of an organic–inorganic interpenetration network. The ^{29}Si NMR spectra demonstrated that the polymerization degree of the aluminosilicate and the chemical environment around the $[SiO_4]^{4-}$ tetrahedra were affected by the introduction of PAV into the geopolymer matrix. An appropriate PVA content incorporated into slag-based geopolymer led to the increased generation of layered-group (Q^3) tetrahedra, indicating an increase in the polymerization degree of aluminosilicate tetrahedra.
4. BSE analysis showed that PVA uniformly dispersed in the inorganic matrix without an obvious interface under dry curing conditions. Due to the influence of the pressure-mixing process, the reaction product bound more closely to the unreacted slag.

Author Contributions: Writing—review and editing, X.X.; Conceptualization and supervision, W.X.; formal analysis, investigation, and data curation, G.Z.; methodology, X.W. All authors have read and agreed to the published version of the manuscript.

Funding: This research was funded by National Natural Science Foundation of China, grant number 52178208; the Fundamental Research Funds for the Central Universities, grant number 2023ZYGXZR031; the Natural Science Foundation of Guangdong, China, grant number 2021 A1515010613; and the Science and Technology Program of Guangzhou, China, grant number 202201010738.

Data Availability Statement: Data are contained within the article.

Acknowledgments: The authors acknowledge technical support given by the Analysis and Test Center, South China University of Technology.

Conflicts of Interest: Author Xilian Wen was employed by the company Guangzhou Residential Construction Development Co., Ltd. The remaining authors declare that the research was conducted in the absence of any commercial or financial relationships that could be construed as a potential conflict of interest.

References

1. Duxson, P.; Thaumaturge, C. Fracture toughness of geopolymer concretes reinforced with basalt fibers. *Cem. Concr. Compos.* **2005**, *27*, 49–54.
2. Akarken, G.; Cengiz, U. Fabrication and characterization of metakaolin-based fiber reinforced fire resistant geopolymer. *Appl. Clay Sci.* **2023**, *232*, 106786. [CrossRef]
3. Gou, H.; Rupasinghe, M.; Sofi, M.; Sharma, R.; Ranzi, G.; Mendis, P.; Zhang, Z. A Review on Cementitious and Geopolymer Composites with Lithium Slag Incorporation. *Materials* **2024**, *17*, 142. [CrossRef]
4. Meftah, M.; Oueslati, W.; Chorfi, N.; Amara, A.B. Intrinsic parameters involved in the synthesis of metakaolin based geopolymer: Microstructure analysis. *J. Alloys Compd.* **2016**, *688*, 946–956. [CrossRef]
5. Davidovits, J. Geopolymer: Inorganic polymeric new materials. *J. Therm. Anal. Calorim.* **1994**, *37*, 1633–1656. [CrossRef]
6. Astutiningsih, S.; Liu, Y.N. Geopolymerisation of Australian slag with effective dissolution by the alkali. In Proceedings of the World Congress Geopolymer 2005, Saint Quentin, France, 28 June–1 July 2005.
7. Kim, D.; Lai, H.T.; Chilingar, G.V.; Yen, T.F. Geopolymer formation and its unique properties. *Environ. Geol.* **2006**, *51*, 103–111. [CrossRef]

8. Yuan, J.K.; He, P.G.; Zhang, P.F.; Jia, D.C.; Cai, D.L.; Yang, Z.H.; Duan, X.M.; Wang, S.J.; Zhou, Y. Novel geopolymer based composites reinforced with stainless steel mesh and chromium powder. *Constr. Build. Mater.* **2017**, *150*, 89–94. [CrossRef]
9. Lin, T.S.; Jia, D.C.; Wang, M.R.; He, P.G.; Lang, D.F. Effects of fibre content on mechanical properties and fracture behaviour of short carbon fibre reinforced geopolymer matrix composites. *Bull. Mater. Sci.* **2009**, *32*, 77–81. [CrossRef]
10. Özkan, S.; Demir, F. The hybrid effects of PVA fiber and basalt fiber on mechanical performance of cost effective hybrid cementitious composites. *Constr. Build. Mater.* **2020**, *263*, 120564. [CrossRef]
11. Li, W.; Xu, J. Mechanical properties of basalt fiber reinforced geopolymeric concrete under impact loading. *Mat. Sci. Eng.* **2008**, *505*, 178–186. [CrossRef]
12. Birchall, J.D.; Howard, A.J.; Kendall, K. Flexural strength and porosity of cements. *Nature* **1981**, *289*, 388–390. [CrossRef]
13. Birchall, J.D.; Kendall, K.; Howard, A.J. Cementitious Product 1981. European Patent Office EP0021682, 6 May 1981.
14. Birchall, J.D.; Howard, A.J.; Kendall, K. A cement spring. *J. Mater. Sci. Lett.* **1982**, *1*, 125–126. [CrossRef]
15. Oludele, O.P.; Kriven, W.M.; Young, J.F. Microstructural and microchemical characterization of a calcium aluminate-polymer composite (MDF cement). *J. Am. Ceram. Soc.* **1991**, *74*, 1928–1933.
16. Bortzmeyer, D.; Frouin, L.; Montardi, Y.; Orange, G. Microstructure and mechanical properties of macro defect-free cements. *J. Mater. Sci.* **1995**, *30*, 4138–4144. [CrossRef]
17. Rodger, S.A.; Brooks, S.A.; Sinclair, W. High strength cement pastes part 2: Reactions during setting. *J. Mater. Sci.* **1985**, *20*, 2853–2860. [CrossRef]
18. Tan, L.S.; McHugh, A.J.; Gulgun, M.A.; Kriven, W.M. Evolution of mechano-chemistry and microstructure of a calcium aluminate-polymer composite: Part II. Mixing rate effects. *J. Mater. Res.* **1996**, *11*, 1739–1747. [CrossRef]
19. Catauroa, M.; Papale, F.; Lamanna, G.; Bollino, F. Geopolymer/PEG hybrid materials synthesis and investigation of the polymer influence on microstructure and mechanical behavior. *Mater. Res.* **2015**, *18*, 698–705. [CrossRef]
20. Mikhailova, O.; Rovnaník, P. Effect of polyethylene glycol addition on metakaolin-based geopolymer. *Procedia Eng.* **2016**, *151*, 222–228. [CrossRef]
21. Rai, U.S.; Singh, R.K. Effect of polyacrylamide on the different properties of cement and mortar. *Mater. Sci. Eng. A* **2005**, *392*, 42–50. [CrossRef]
22. Manzur, T.; Iffat, S.; Noor, M.A. Efficiency of sodium polyacrylate to improve durability of concrete under adverse curing condition. *Adv. Mater. Sci. Eng.* **2015**, *8*, 685785. [CrossRef]
23. Chang, J.J. A study on the setting characteristics of sodium silicate-activated slag pastes. *Cem. Concr. Res.* **2003**, *33*, 1005–1011. [CrossRef]
24. *GB/T18046-2008*; Granulated Blast Furnace Slag Powder for Cement and Concrete. Standards Press of China: Beijing, China, 2008.
25. Pan, H.; She, W.; Zuo, W.; Zhou, Y.; Liu, J. Hierarchical toughening of a biomimetic bulk cement composite. *ACS Appl. Mater. Interfaces* **2020**, *12*, 53297–53309. [CrossRef]
26. Bonapasta, A.A. Cross-Linking of Poly(vinyl alcohol) Chains by Al Ions in Macro-Defect-Free Cements: A Theoretical Study. *Chem. Mater.* **2000**, *12*, 738–743. [CrossRef]
27. Ferrante, F.; Bertini, M.; Ferlito, C.; Lisuzzo, L.; Lazzara, G.; Duca, D. A computational and experimental investigation of halloysite silicic surface modifications after alkaline treatment. *Appl. Clay Sci.* **2023**, *232*, 106813. [CrossRef]
28. Andersen, M.D.; Jakobsen, H.J.; Skibsted, J. Characterization of white Portland cement hydration and the C-S-H structure in the presence of sodium aluminate by ^{27}Al and ^{29}Si MAS NMR spectroscopy. *Cem. Concr. Res.* **2004**, *34*, 857–868. [CrossRef]
29. Wang, L.; He, Z. Revealing C-S-H polymerization mechanism based on infrared and nuclear magnetic resonance technology. *J. Build. Mater.* **2011**, *14*, 447–451.
30. Aronne, A.; Esposito, S.; Ferone, C.; Pansini, M.; Pernice, P. FTIR study of the thermal transformation of barium exchanged zeolite A to celsian. *J. Mater. Chem.* **2002**, *12*, 3039–3045. [CrossRef]
31. Catauro, M.; Raucci, M.G.; De Gaetano, F.; Marotta, A. Sol–gel synthesis, characterization and bioactivity of polycaprolactone/ SiO_2 hybrid material. *J. Mater. Sci.* **2003**, *38*, 3097–3102. [CrossRef]
32. Zhang, Y.; Sun, W.; Li, Z. Infrared spectroscopy study of structural nature of geopolymeric products. *J. Wuhan Univ. Technol.-Mater. Sci. Ed.* **2008**, *23*, 522–527. [CrossRef]
33. Barbosa, V.F.F.; MacKenzie, K.J.D. Thermal behavior of inorganic geopolymer and composites derived from sodium polysialate. *Mater. Res. Bull.* **2003**, *38*, 319–331. [CrossRef]
34. Chen, X.; Zhu, G. Effect of organic polymers on the properties of slag-based geopolymers. *Constr. Build. Mater.* **2018**, *167*, 216–224. [CrossRef]
35. Ma, B.G.; Hai, N.L.; Xiang, G.L. Influence of nano-TiO_2 on physical and hydration characteristics of fly ash-cement systems. *Constr. Build. Mater.* **2016**, *122*, 242–253. [CrossRef]

Disclaimer/Publisher's Note: The statements, opinions and data contained in all publications are solely those of the individual author(s) and contributor(s) and not of MDPI and/or the editor(s). MDPI and/or the editor(s) disclaim responsibility for any injury to people or property resulting from any ideas, methods, instructions or products referred to in the content.

Article

Study on the Preparation and Properties of Vegetation Lightweight Porous Concrete

Qingyu Cao [1], Juncheng Zhou [2], Weiting Xu [3,*] and Xiongzhou Yuan [4,*]

[1] MMC Group, Central Research Institute of Building and Construction, Beijing 10088, China; achero@126.com
[2] School of Civil Engineering, Inner Mongolia University of Science and Technology, Baotou 014010, China; 17692455153@163.com
[3] School of Materials Science and Engineering, South China University of Technology, Guangzhou 510641, China
[4] School of Traffic and Environment, Shenzhen Institute of Information Technology, Shenzhen 518172, China
* Correspondence: xuweiting@scut.edu.cn (W.X.); yuanxz@sziit.edu.cn (X.Y.)

Abstract: The objective of this study is to formulate vegetated light porous concrete (VLPC) through the utilization of various cementing materials, the design of porosity, and the incorporation of mineral additives. Subsequently, the study aims to assess and analyze key properties, including the bulk density, permeability coefficient, mechanical characteristics, and alkalinity. The findings indicate a linear decrease in the volume weight of VLPC as the designed porosity increases. While higher design porosity elevates the permeability coefficient, the measured effective porosity closely aligns with the design values. The examined VLPC exhibits a peak compressive strength of 17.7 MPa and a maximum bending strength of 2.1 MPa after 28 days. Notably, an escalation in porosity corresponds to a decrease in both the compressive and the bending strength of VLPC. Introducing mineral additives, particularly silicon powder, is shown to be effective in enhancing the strength of VLPC. Furthermore, substituting slag sulfonate cement for ordinary cement significantly diminishes the alkalinity of VLPC, resulting in a pH below 8.5 at 28 days. Mineral additives also contribute to a reduction in the pH of concrete. Among them, silica fume, fly ash, fly ash + slag powder, and slag powder exhibit a progressively enhanced alkaline reduction effect.

Keywords: porosity; mechanical property; alkalinity; permeability coefficient

Citation: Cao, Q.; Zhou, J.; Xu, W.; Yuan, X. Study on the Preparation and Properties of Vegetation Lightweight Porous Concrete. *Materials* **2024**, *17*, 251. https://doi.org/10.3390/ma17010251

Academic Editor: Carlos Leiva

Received: 20 November 2023
Revised: 27 December 2023
Accepted: 28 December 2023
Published: 3 January 2024

Copyright: © 2024 by the authors. Licensee MDPI, Basel, Switzerland. This article is an open access article distributed under the terms and conditions of the Creative Commons Attribution (CC BY) license (https:// creativecommons.org/licenses/by/ 4.0/).

1. Introduction

In recent years, vegetation lightweight porous concrete (VLPC) has emerged as an innovative integration of traditional concrete technology with the advantages of horticulture [1,2]. VLPC boasts a considerable pore volume and a distinctive pore structure [3–6], resulting in enhanced water permeability [7], efficient heat dissipation capacity [8], and notable sound absorption capability [9]. These inherent characteristics render VLPC highly valuable for mitigating urban waterlogging [10], minimizing noise pollution [11], and fostering urban ecological development. Consequently, VLPC plays a pivotal role in the implementation of the concept of sponge cities [3]. By integrating plants within a porous concrete medium, VLPC enables the unhindered movement of water, air, soil, and roots. This environmentally friendly material finds frequent application in valley slopes and hilly areas, where it serves to enhance landscaping and to mitigate the risk of slope landslides. Nevertheless, in practical applications, the predominant cement used is typically ordinary Portland cement. This type of cement generates a substantial amount of the hydration product $Ca(OH)_2$ during the hydration process, creating an alkaline environment that is unfavorable for vegetation growth [12,13]. The optimal pH range for plant growth is between 8 and 10, whereas the pH value of ordinary concrete is approximately 13—clearly unsuitable for the formulation of VLPC [14].

Numerous researchers have undertaken comprehensive studies with the objective of creating an internal environment within VLPC that promotes vegetation growth [15–18]. Li et al. [19] observed that, at a water–cement ratio of 0.26 and a targeted porosity of 26%, both the pH value of the soil and the concrete porosity reached their lowest levels. This condition was identified as providing the most optimal environment for vegetation growth. These findings underscore the significance of considering the porosity and pH value as critical factors influencing the growth and development of vegetation in VLPC. Peng et al. [20] conducted a comparative analysis of the permeability of concrete incorporating fly ash and silica fume. Their findings revealed that the addition of fly ash or silica fume led to a reduction in permeability while concurrently improving the strength of the concrete. Consequently, the determination of the optimal quantity of fly ash or silica fume in the manufacturing process of VLPC should be guided by primary performance criteria. Additionally, some scholars have noted that altering the raw material composition has a positive impact on enhancing the alkaline environment within the concrete. Yang et al. [21] investigated the evolution process and mechanism of the engineering performance of vegetated concrete under atmospheric freeze–thaw (F-T) test conditions. The findings indicated a decline in the acoustic wave velocity and cohesive forces, coupled with an increase in the permeability coefficient of vegetated concrete due to the effects of F-T action. The reduction in cohesive force was closely linked to an overall decrease in the content of the gelling hydration products in the vegetated concrete. Additionally, the diminished nutrient retention capacity of vegetated concrete was primarily associated with the disintegration and fragmentation of larger aggregates induced by F-T action. Cheng et al. [13] devised 10 sets of porous concrete specimens using mixed ratio columns to examine the influence of various parameters, including fly ash and silica fume contents, on the mechanical properties, porosity, and pH value of porous concrete. The outcomes of the study revealed that the compressive strength and porosity of the porous concrete, measured after 28 days, were at least 13 MPa and 21%, respectively. This suggests that the porous concrete exhibits suitability for slope protection in practical engineering applications. Additionally, all grass species thrived in the eco-modified soil and porous concrete environment throughout the observation period, achieving a vegetation coverage rate of 96% or higher. It was also observed that the root system permeated the pores of the porous concrete, reaching the bottom after 8 weeks. Kim et al. [22] conducted a study evaluating the vegetation growth performance of porous vegetated lightweight porous concrete (VLPC) blocks prepared with blast furnace slag cement. After placing the test blocks in the field, plant growth was monitored, and seeds began to germinate a week after sowing. Remarkably, after 6 weeks, the plant length exceeded 300 mm, and the average coverage reached an impressive 90%. These findings unequivocally demonstrate that VLPC, prepared with blast furnace slag cement, is well-suited for environmental remediation projects. Moreover, the selection of aggregates and cement substitutes was identified as a significant factor influencing the vegetation growth performance of VLPC. Wang et al. [1] found that soaking in ferrous sulfate solution for 6 days, oxalic acid solution for 10 days, and water for 26 days can reduce the pH value of the internal pores of concrete to about 8, indicating that the alkali reduction effect can be achieved in a short time by using an alkali reduction solution, and the other properties of vegetation concrete are not significantly affected. Bao et al. [8] pioneered the development and investigation of vegetation pervious concrete with low alkalinity. The reduction in concrete alkalinity was achieved through the addition of admixtures to the cement slurry. The study utilized XRD (X-ray diffraction) and compression tests to examine the impact of the admixture content on the concrete basicity and compressive strength. The results indicated a decrease in the alkalinity of the cement samples with increase in the admixture content and successful vegetation growth on the pervious concrete. Moreover, by elevating the admixture content to approximately 3.6%, the compressive strength of the pervious concrete exceeded 25 MPa.

Building upon these findings, the present study undertook the preparation of various types of plant-growing lightweight porous concrete. To cater to the requirements of vegetation growth, essential performance indicators, including the bulk density, strength, alkalinity, porosity, and water permeability coefficient, were meticulously examined. The objective was to furnish a scientific foundation and technical support for the application of this material in construction engineering, urban greening, as well as energy-saving and environmental protection initiatives.

2. Materials and Methods

2.1. Materials

P.II 52.5 grade ordinary Portland cement (OPC) (Onoda Cement Factory, Dalian, China), meeting the Chinese standard GB/T 175-2007 [23], and 52.5 grade slag sulphoaluminate cement (SSC) (Polar bear Co., Tangshan, China), meeting the Chinese standard GB/T 20472-2006 [24], were chosen for the experiment. The specific surface areas of these cements were 392 m^2/kg and 450 m^2/kg, respectively. The coarse aggregate utilized was shale ceramic granule 900 grade lightweight aggregate, with a density of 960 kg/m^3 and a particle size ranging from 10 to 15 mm. Fly ash (Datang Tongzhou thermal power plant, Tangshan, China) was employed as low-calcium work-grade fly ash (FA) with a density of 2.13 g/cm^3, a fineness of 6% (45 μm sieve-screened), and a water demand ratio of 0.95. The slag powder used was S95 granulated blast furnace slag (GBFS) (Douhe power plant, Tangshan, China), characterized by a density of 2.86 g/cm^3, a specific surface area of 590 m^2/kg, a mobility ratio of 97, and a 28-day activity index of 104. The chemical additive chosen was JM-2 polycarboxylic (Sobute New Materials Co., Nanjing, Jiangsu, China) acid water-reducing agent, achieving a water reduction rate of 36.7%. The chemical compositions of the cement, FA, GBFS, and silica fume (SF) (Elkem Materials Co., Guangzhou, China) are detailed in Table 1.

Table 1. Chemical composition of SSC, OPC, FA, SF and GBFS.

Oxides%	CaO	SiO$_2$	Al$_2$O$_3$	Fe$_2$O$_3$	MgO	SO$_3$	TiO$_2$	LOI
P.II 52.5	64.50	22.04	4.76	3.10	0.92	1.90	/	1.01
SSC52.5	39.25	26.95	13.21	0.41	8.51	7.84	0.53	0.31
FA	3.36	57.74	27.08	6.34	1.11	0.18	/	3.61
SF	0.4	91.05	1.73	0.91	0.78	0.27	/	1.02
GBFS	37.19	31.77	15.36	0.63	10.15	1.16	/	1.02

2.2. Mix Design

The preparation of vegetative lightweight porous concrete relies on effectively enveloping the coarse aggregate with cementitious material to create uninterrupted voids. However, conventional concrete proportion design methods do not meet the necessary criteria for void formation in this specific type of concrete. Therefore, it is essential to devise novel and effective design principles and techniques for the proportioning of lightweight and porous vegetative concrete.

The mix ratio of vegetated lightweight porous concrete is influenced by several crucial factors, including the water–cement ratio ($R_{W/C}$), the design porosity (R_{void}), the density (ρ), and the aggregate porosity (v_c). The specific calculations can be broken down into the following steps:

(1) Calculation of the amount of coarse aggregate per unit cubic meter.

$$W_G = \alpha \cdot \rho_G \ (kg/m^3), \tag{1}$$

where α represents the correction coefficient, set at 0.98; ρ_G signifies the compact packing density of coarse aggregate, measured in kilograms per cubic meter (kg/m^3); and W_G denotes the quantity of coarse aggregate per unit cubic meter, expressed in kilograms per cubic meter (kg/m^3).

(2) Calculation of slurry volume of cementitious materials.

$$V_P = 1000 - \alpha \cdot 10 \cdot (100 - v_c) - R_{void} \cdot 10 \ (L/m^3), \quad (2)$$

where V_P denotes the volume of cementitious material (kg/m³); v_c denotes the dense pile porosity of coarse aggregate (%); and R_{void} denotes the design porosity (%).

(3) Calculation of cement and water consumption per cubic meter.

$$W_C = \frac{V_p}{R_{W/C} + \frac{1}{\rho_C}} \quad (3)$$

$$W_W = W_C \cdot R_{W/C}, \quad (4)$$

where W_W denotes the amount of cement per unit cubic metre (kg/m³); $R_{W/C}$ denotes the water-cement ratio; ρ_C denotes the density of cement (kg/m³); and W_W denotes the amount of water used per unit cubic metre (kg/m³).

When incorporating fly ash, slag powder, silica fume, and other mineral admixtures, the dosage calculations are based on volume conversion corresponding to the volume of the cementitious material slurry. This involves following the aforementioned steps to calculate the dosage for each admixture. Typically, when the admixture dosage is small, its volume is not factored into the total volume of the slurry. The specific calculations for various groups are detailed in Table 2.

Table 2. Mix proportion (kg/m³).

Number	Design Porosity (%)	Cement	Aggregate	Water	FA	GBFS	SF	JM-2	Water-Cement Ratio
C1	20	406	524	105.6				2.18	0.26
C2	23	356	524	92.6				1.79	0.26
C3	25	322	524	83.7				1.73	0.26
C4	27	288	524	74.9				1.42	0.26
C5	30	234	524	60.8				1.05	0.26
CF	25	170	513	79.2	113.0			1.42	0.28
CG	25	182	513	88.2		122.0		1.52	0.29
CS	25	325	513	79.7			6.6	1.66	0.24
CD	25	177	513	82.6	59.0	59.0		1.48	0.28
S	25	335	524	83.7				1.73	0.26

Note: C stands for common silicate cement; S stands for slag sulfur–aluminate cement. CF, CG, CS, and CD represent samples with single-doped FA, single-doped GBFS, single-doped SF, and double-doped FA and GBFS, respectively.

2.3. Specimen Preparation and Maintenance

In contrast to ordinary concrete, the fluidity of VLPC is significantly reduced, attributed to its lower cement content, absence of fine aggregate, and the presence of a substantial amount of coarse aggregate. The internal friction between the slurry and coarse aggregate, along with the coarse aggregate itself, is notably high. This results in the slurry being prone to forming clusters, making it not only challenging to mix but also difficult to establish a continuous and cohesive structure. To ensure the comprehensive and effective coverage of the coarse aggregate surface by the cementing material slurry, a novel stirring process is introduced in the test, as illustrated in Figure 1. The mixing process involves several steps: initially, cement, mineral admixture, and one-third of the coarse aggregate are mixed and stirred for 15 s. Subsequently, half of the water and admixture are added, and the mixture is stirred for 30 s. Following this, a third of the coarse aggregate, along with a quarter of the water and admixture, is introduced into the concrete, and mixing is continued for 30 s. Lastly, the remaining coarse aggregate, water, and admixture are added, and the entire mixture is mixed and stirred for 60 s. The resulting concrete is then poured to form the desired structure. Concrete specimens of varying heights (100 mm × 100 mm × 100 mm and 100 mm × 100 mm × 400 mm) were prepared based on the mix ratios in Table 2. After 24 h of casting, the concrete specimens were demolded, followed by a subsequent 28-day curing process in a standard curing room maintained at a temperature of 20 ± 2 °C and a relative humidity of ≥95%.

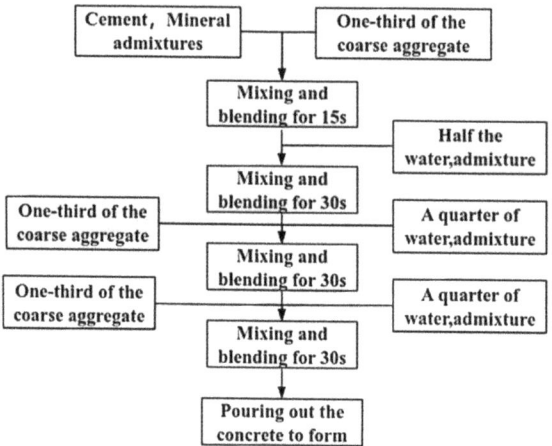

Figure 1. VLPC mixing process.

2.4. Test Methods and Procedures

2.4.1. Volumetric Weight of Vegetated Lightweight Porous Concrete

Given the substantial differences in composition and structure between vegetative lightweight porous concrete and ordinary concrete, it is anticipated that their bulk densities will also vary significantly. The testing procedure followed the mixed bulk density test used for ordinary concrete, and the concrete was sealed before the test, focusing on Group C concrete with differing porosities to measure the mixed bulk density.

2.4.2. Porosity and Permeability Coefficient

The porosity of vegetative lightweight porous concrete (VLPC) was assessed in accordance with the specifications outlined in CJJ/T253-2016 [25]. Simultaneously, the permeability coefficient of VLPC was determined using the constant head method, following the guidelines provided in CJJ/T135-2009 [26], as depicted in Figure 2.

Porosity is determined by measuring the change in water level before and after immersing the sample to ascertain the drainage volume. The formula for calculating the porosity is provided in Equation (5).

$$P = \frac{V - S \times (H_2 - H_1)}{V} \times 100\%, \qquad (5)$$

where V denotes the exterior volume of the specimen (cm^3); S denotes the cross-sectional area of the container (cm^2); H_1 indicates the initial liquid level height (cm); and H_2 indicates the liquid surface height after adding the specimen (cm). When the difference between the maximum value, the minimum value, and the intermediate value exceeds ±0.5%, the sample is re-sampled for the test, and the arithmetic average of the three test results is taken as the final measured value.

The formula for calculating the permeability coefficient is shown in Equation (6):

$$K_T = \frac{QD}{AH(t_2 - t_1)}, \qquad (6)$$

where K indicates the coefficient of water permeability (cm/s) at water temperature $T\ ^\circ C$; Q denotes the amount of water (cm^3) that passes through the concrete from time t_1 to t_2; D denotes the thickness of the macroporous concrete specimen (mm); A denotes the contact area of the macroporous concrete specimen (cm^2); and H denotes the specified head height (cm). It should be noted that due to the selection of light aggregate with large particle size to prepare VLPC in this test, its water permeability is very good, and the equilibrium head is very small, and it is difficult to reach the specified head. Therefore, the method of balancing the head instead of the specified head is used to calculate the water permeability coefficient.

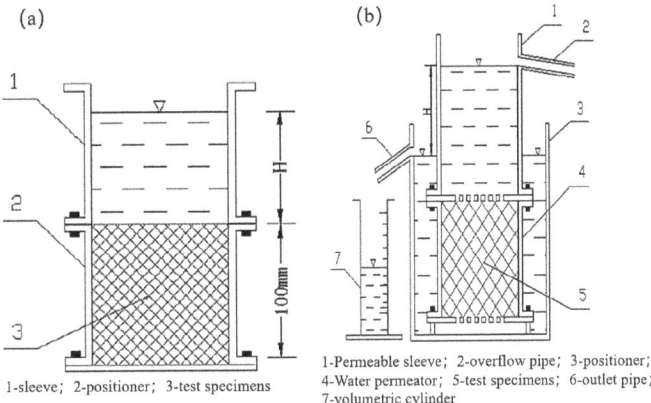

Figure 2. Measuring instrument for porosity and permeability coefficient. (**a**) Porosity measuring instrument; (**b**) permeability coefficient measuring instrument.

2.4.3. Mechanical Properties

The strength test for VLPC follows the guidelines specified in GB/T50081-2019 [27]. The testing instrument employed is a 300-ton electro-hydraulic servo-pressure testing machine, with a control loading rate of 0.2 MPa/s. The test ages selected are 7 and 28 days, respectively. As the test blocks used in this study are non-standard specimens, the test flexural strength is adjusted by multiplying it by the conversion coefficient of 0.85, and the test compressive strength is adjusted by multiplying it by the conversion coefficient of 0.95 [27]. The average compressive strength of the three concrete test blocks is considered as the strength value for the specimen group, with the requirement that the deviation between the individual strengths and the average strength should not exceed 15%.

2.4.4. Alkalinity

Due to the inhibitory effect of high alkalinity on seed germination and root growth, the alkalinity of concrete is mitigated by substituting cement types with mineral admixtures [28,29]. The test was conducted following the procedures recommended in IS: 3025 (Part II) -1983, and the sample for the alkalinity test was prepared in accordance with the procedure outlined in ref. [30]. For sample preparation, after reaching the specified age, the sample is extracted from the standard curing chamber and subjected to crushing and grinding, followed by passage through a 0.08 mm sieve. Subsequently, 10 g of the powdered sample is weighed, mixed with 10 times distilled water, and the bottle is sealed to prevent carbonization. After allowing it to sit for 2 h, the mixture is filtered using filter paper. The alkalinity of the concrete is then determined based on the pH value of the filtered solution. The experimental setup for the pH test is illustrated in Figure 3.

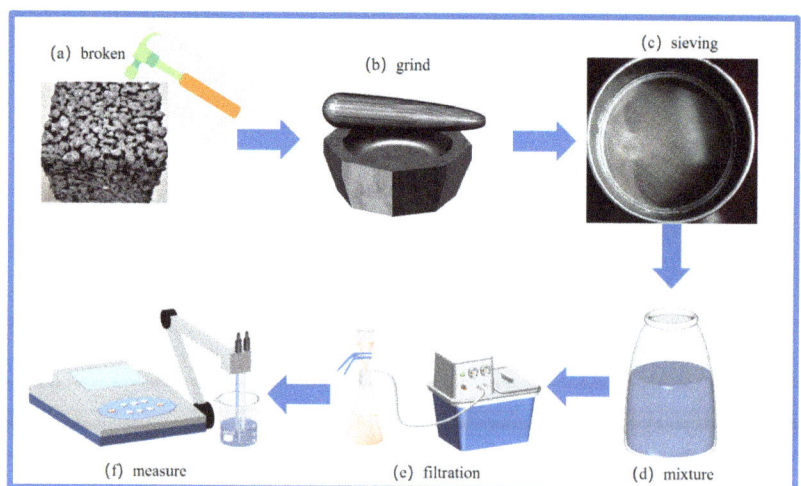

Figure 3. pH Testing. (**a**) Broken sample; (**b**) Grinding sample; (**c**) Sample sieving; (**d**) Mix with distilled water; (**e**) Filter mixture; (**f**) pH measurement.

3. Results

3.1. Bulk Weight of Concrete

The results of the volumetric weight determination for mixes with different design porosities are presented in Figure 4. From the figure, it is evident that the bulk weight of fresh VLPC in Group C1 ranges from 1200 kg/m^3 to 1300 kg/m^3, approximately half of the bulk weight of ordinary concrete (2200 kg/m^3 to 2500 kg/m^3), categorizing VLPC as a lightweight concrete. With the same mixing ratio, an increase in the design porosity correlates with a decrease in the mix weight, demonstrating a predominantly linear relationship between them. In essence, an elevation in the design porosity results in a reduced mass of the active substance within the same volume of concrete, consequently leading to a decline in the mix's bulk weight.

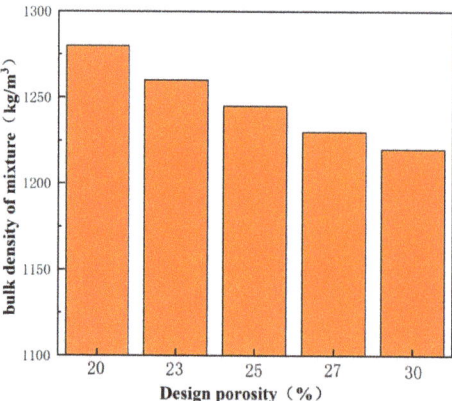

Figure 4. Concrete bulk weight test results.

3.2. Porosity and Permeability Coefficient

The effective porosity and water permeability coefficient of five test blocks with varying design porosities in Group C were determined. The corresponding relationship between the design porosity and the effective porosity is depicted in Figure 5, while the test results for the water permeability coefficients are illustrated in Figure 6.

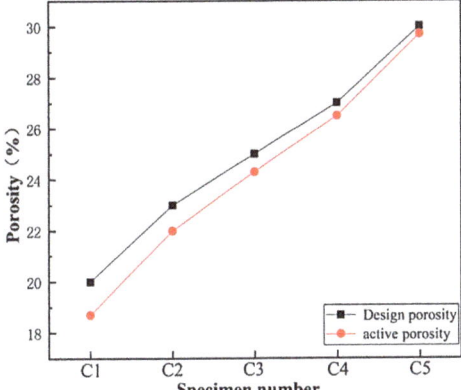

Figure 5. Effective porosity test results.

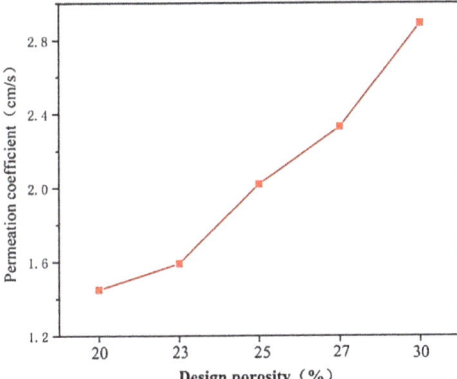

Figure 6. Permeability coefficient test results.

From the observations in Figure 5, it is evident that the measured effective porosity of the lightweight porous concrete is smaller than that of the design porosity. This phenomenon is attributed to the occlusion of some pores during the concrete molding process, hindering the formation of continuous or semi-continuous effective porosity. Furthermore, the effective porosity of porous concrete is somewhat influenced by the pressure during molding and the particle size of the coarse aggregate. In this study, a single primary distribution particle size of 10 mm to 20 mm was utilized for the lightweight aggregate, and the disparity between the measured effective porosity and the design porosity was not significant. When the reserved porosity in the design is larger, it is more likely to form continuous or semi-continuous effective porosity, resulting in fewer invalid pores that are completely occluded. Consequently, the larger the design porosity, the closer the measured effective porosity is to the design value. This trend is highly favorable for porous concrete. During the molding process, when the mix of porous concrete is subjected to external forces, the larger reserved pore space can accommodate the movement of the cementitious material slurry wrapped with coarse aggregate. This reduces the formation of completely occluded pores, gradually bringing the measured effective porosity closer to the design porosity.

As depicted in Figure 6, the water permeability coefficient of VLPC increases with the rise in design porosity. This is attributed to the larger design porosity, signifying a smaller amount of cementitious material slurry and a reduced volume of wrapped coarse aggregate surface. As a result, the formation of occluded pores inside the VLPC is less likely, and continuous pores naturally increase, enhancing the water permeability performance.

3.3. Mechanical Properties

3.3.1. Effect of Porosity on Mechanical Properties

The compressive and flexural strengths of specimens with five different design porosities in Group C1 are compared, as illustrated in Figure 7. From Figure 7a, it is evident that the cubic compressive strength of porous concrete decreases with increasing design porosity at the same proportion. Notably, at the age of 28 days, the compressive strength of porous concrete with 30% porosity decreases by more than 50% compared to that with 20% porosity. This further confirms the significant impact of the pore structure within the porous concrete on macroscopic strength. For porous concrete intended for engineering applications, it is necessary for its 28-day compressive strength to be no less than 10 MPa and the porosity to be not less than 22%. However, based on the trend observed in the figure, there is very limited scope for improving the compressive strength of porous concrete with an increase in the designed porosity. Therefore, in order to meet the engineering design requirements for strength, the design porosity of porous concrete should not be higher than 30%.

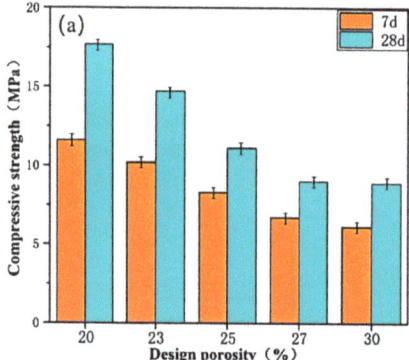
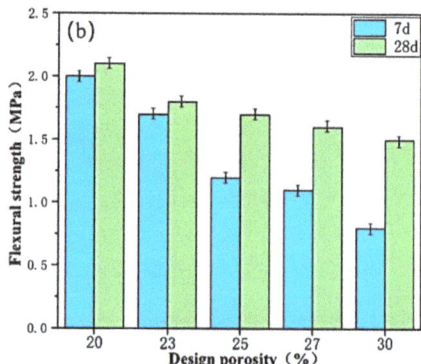

Figure 7. Effect of porosity on compressive and flexural strengths. (**a**) Compressive strength; (**b**) flexural strength.

As observed in Figure 7b, the flexural breaking strength of porous concrete decreases with an increase in the porosity. The higher the porosity, the lower the overall compactness of the porous concrete, leading to a reduction in its flexural performance. When the porosity is low, there are relatively fewer pores in the concrete, and the contact area between the particles is larger. This condition enables effective stress transfer, enhancing the flexural capacity of the concrete. Furthermore, lower porosity implies fewer voids within the concrete, reducing internal weak points and promoting a more uniform and robust structure. This aspect contributes to an enhancement in the flexural strength. As the porosity increases, more voids are present in the concrete, some of which may lead to the formation of concentrated stresses and cracks, thereby diminishing the flexural strength of the concrete. In particular, when the connections between the pores are more pronounced, there is a higher likelihood of cracks being transmitted between the pores, further weakening the overall flexural resistance.

3.3.2. Effect of Mineral Admixtures on Mechanical Properties

The impact of various mineral admixtures on the compressive and flexural strength of VLPC is illustrated in Figure 8. As shown in Figure 8a, both the 7-day and 28-day compressive strengths of VLPC with single doping of GBFS, SF, and double doping of GBFS and FA surpassed that of the C3 group without mineral external dopants. The most notable improvement was observed with single doping of SF, where the 7-day compressive strength increased by 107.1% compared to that of the C3 group. The pozzolanic reaction between $Ca(OH)_2$ and SiO_2 in the SF was expedited by the finer texture of SF. Additionally, the formation of C-S-H resulted from the reaction between the amorphous SiO_2 in SF and calcium hydroxide from cement hydration, thereby enhancing the compressive strength [31]. Liang Fu et al. discovered that the addition of SF to the cement improved the strength and resistance of the concrete against sulphate attack. These findings align with the results presented in this study [32]. The strength values of the single-doped GBFS increased significantly, and the double-doped group CD also experienced some improvement. However, only the compressive strength values of the single-doped FA decreased, aligning with previous literature reports [33,34]. Ibrahim found that adding FA to concrete reduces the early compressive strength of concrete because FA is less active [35]. The compressive strength values in descending order are SF > GBFA > GBFA + FA > Blank > FA.

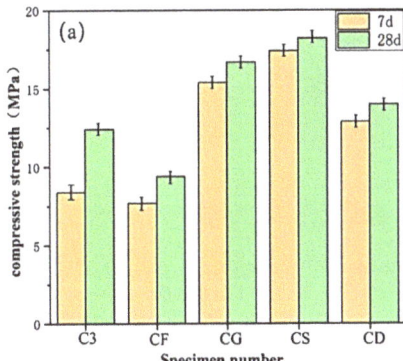
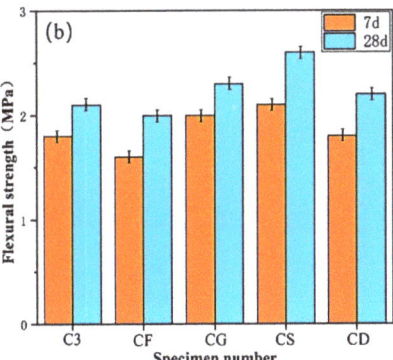

Figure 8. Effect of mineral admixtures on compressive and flexural strengths. (a) Compressive strength; (b) flexural strength.

This indicates that the addition of mineral admixtures plays a significant role in enhancing the compressive strength of VLPC. Similar to ordinary concrete, the introduction of mineral admixtures improves the uniform distribution of the hydration products in space, accelerating the hydration of cementitious materials through its microcrystalline nucleation effect. Moreover, the unique properties of ultrafine micronized powders, such as slag and silica fume, reduce the amount of water used by cementitious materials. Simultaneously, they refine the pore structure, improve its size and morphology, and enhance the density of the cementitious stone. Consequently, these factors collectively contribute to a substantial increase in the strength performance of porous concrete [36]. The crucial aspect lies in the addition of slag and silica fume, along with other ultra-fine micronized powder. This addition facilitates the reduction of the $Ca(OH)_2$ crystal size in the transition zone between cementitious materials and aggregates, weakening the orientation. As a result, the structure of the interface transition layer is significantly improved, enhancing the bond between the cementitious materials and aggregates and inhibiting the destruction of the bond [37].

The impact of mineral external admixtures on the flexural strength of VLPC is illustrated in Figure 8b. The pattern of the effect of mineral external dopants on the flexural strength of VLPC is essentially the same as that observed for the compressive strength. Specifically, the values of the flexural strength are ranked from high to low as SF > GBFA > GBFA + FA > Blank > FA.

3.4. Alkalinity

The pH of different groups of VLPC was measured, and the test results are depicted in Figure 9. After the hydration reaction of normal silicate cement, various compounds are produced, including C-S-H gel, calcium hydroxide, calcium alumina, and hydrated calcium monosulphoaluminate. The concrete environment is primarily alkaline due to the presence of $Ca(OH)_2$, with small amounts of sodium hydroxide and potassium hydroxide also present. Consequently, the $Ca(OH)_2$ content of concrete directly influences the alkalinity of its environment [38].

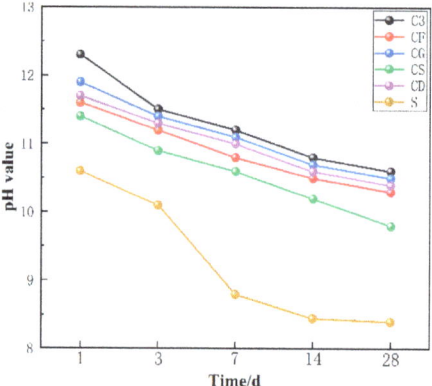

Figure 9. pH value of VLPC at different ages.

Comparing the pH changes of group C3 and group S, the pH of porous concrete prepared with slag sulfoaluminate cement was reduced to less than 8.5 at 28 days, which is 21% lower than that of ordinary silicate cement. This reduction is attributed to the large amount of ettringite generated by the slag sulfoaluminate cement during hydration. The relatively low $Ca(OH)_2$ content contributes to the lower alkalinity of the slag sulfoaluminate cement. Figure 10 illustrates the microstructure of the slag sulfoaluminate cement and silicate cement after 14 days of hydration. It can be observed that the slag sulfoaluminate cement contains a significant amount of ettringite, while the ordinary silicate cement generates a large amount of $Ca(OH)_2$ crystals.

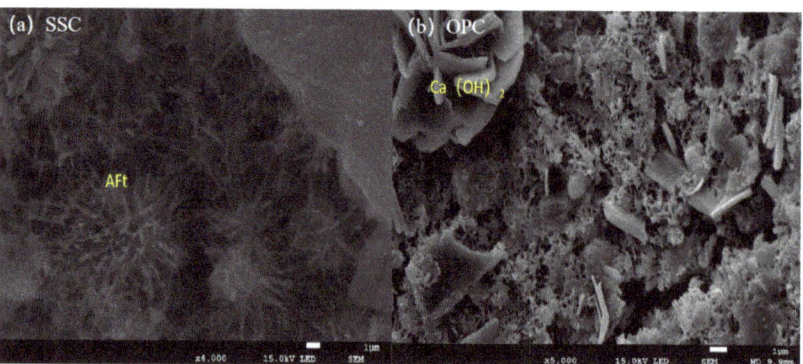

Figure 10. SEM images of SSC hydrated with OPC for 28 days. (**a**) SSC; (**b**) OPC.

As the SiO_2 content in the mineral admixture increases while the CaO content decreases, the water-hardness of SiO_2 leads it to react with Ca^{2+} to generate new hydrated calcium silicate (C-S-H). This reaction consumes some of the Ca^{2+} ions, thereby reducing the $Ca(OH)_2$ content in the concrete. In concrete, $Ca(OH)_2$ is the primary alkaline substance that governs the alkalinity of the concrete. Therefore, as the SiO_2 content increases and the CaO content decreases, the formation of $Ca(OH)_2$ in the concrete decreases, consequently lowering the pH of the internal voids of the concrete. Specifically, the CaO content in silica fume, fly ash, and slag powder gradually increases, while the SiO_2 content gradually decreases. Therefore, the increase in SiO_2 and the decrease in CaO when using these mineral admixtures result in a reduced amount of $Ca(OH)_2$ in the concrete. This ultimately leads to a decrease in the pH of the internal voids of the concrete [39]. This is because the reaction of SiO_2 with $Ca(OH)_2$ reduces the solubility of $Ca(OH)_2$, thereby affecting the alkaline environment of concrete. The pH of concrete can be adjusted by controlling the composition and content of mineral admixtures.

The presence of more and larger internal voids in VLPC allows for increased air contact compared to normal concrete. This heightened exposure to air increases the likelihood of carbonation, resulting in a lower pH for porous concrete in comparison to plain concrete. The pH of the internal voids of porous concrete typically decreases to below 10.5 at 28 days. Furthermore, the pH of the internal voids of VLPC decreases further with the incorporation of various mineral admixtures. Among them, silica fume has the most significant effect, reducing its 28-day pH to below 10, sufficiently meeting the growth requirements of some plants. Additionally, other external admixtures can effectively reduce the pH of porous concrete. Based on the degree of pH reduction observed in this study, the order is SF > FA > FA + GBFS > GBFS.

These findings show that the pH value of the interstitials in porous concrete can be effectively reduced by incorporating various mineral admixtures and using low-alkali cement to create a suitable environment for plant growth. In addition, the pH reduction of different admixtures is different, emphasizing the need for careful consideration and reasonable selection of admixtures according to the specific requirements.

4. Discussion

VLPC employs lightweight aggregate as the coarse aggregate and lacks fine aggregate, resulting in a density ranging from 1200 kg/m³ to 1300 kg/m³. This is approximately half of the density of ordinary concrete (2200 kg/m³ to 2500 kg/m³), enhancing the construction convenience, improving efficiency, and saving materials to a certain extent.

The measured effective porosity of lightweight porous concrete is typically slightly smaller than the design porosity. However, as the design porosity increases, the measured effective porosity gradually approaches the design value. This indicates that the incorporation of larger pre-reserved pores during the molding process assists in forming continuous or semi-continuous effective pores, reducing the occurrence of completely occluded pores. Additionally, it enhances the water permeability of the concrete, which is highly advantageous for porous concrete.

The strength of VLPC decreases with increasing porosity, highlighting the need for a balanced relationship between the porosity and the mechanical properties in the design of porous concrete structures. To ensure sufficient strength, it is essential to control the porosity within an appropriate range, ensuring that the mechanical properties of the concrete meet the design requirements.

The incorporation of mineral admixtures is crucial for enhancing the strength of VLPC. Both slag powder and silica fume contribute significantly to early and mid-late strength development. In cases where economic considerations come into play, a double mix with an appropriate proportion of fly ash, replacing slag powder, can be employed to achieve strength improvement. However, it is not advisable to use fly ash alone. This is because, given that the early strength of vegetative lightweight porous concrete is not inherently high, the addition of fly ash could further reduce its early strength, presenting challenges in the construction of vegetative lightweight porous concrete.

The inclusion of mineral admixtures can variably decrease the alkalinity of VLPC. Higher SiO_2 content in minerals leads to reduced $Ca(OH)_2$ generation, resulting in a lower pH value. When SSC is utilized, the alkalinity of VLPC is notably diminished compared to ordinary concrete, primarily due to its main hydration product being dominated by calcium alumina with a limited $Ca(OH)_2$ content.

5. Limitations and Prospects

5.1. Limitations

In our study, we employed a constrained set of materials, pore size designs, and mineral additives to fabricate VLPC. However, this restricted selection may not fully encompass the potential variations and combinations that could optimize the overall performance of VLPC. Subsequent research and experimentation involving a broader range of materials are necessary to comprehensively explore and enhance its properties.

Our experiments were carried out in controlled laboratory conditions, neglecting the potential impact of long-term exposure to environmental factors or aging. In real-world applications, VLPC may face various weathering conditions, including temperature fluctuations, moisture, and chemical exposure, which could influence its performance and durability. Future studies should include field tests or accelerated aging experiments to evaluate the long-term behavior of VLPC.

5.2. Prospects

VLPC holds significant potential for diverse applications. It proves beneficial in road and pavement construction, green roofs, retaining walls, and landscaping projects. The porous structure of VLPC facilitates water infiltration, addressing urban stormwater runoff issues, and contributing to the enhanced ecological balance of urban environments.

Future research can concentrate on broadening the spectrum of cementitious materials and mineral additives employed in VLPC production. Through the exploration of diverse combinations, researchers aim to enhance the mechanical properties, improve durability, and reduce the costs associated with the manufacturing process.

While VLPC exhibits promise, further studies and practical applications are essential to validate its performance in diverse environments and construction scenarios. Collaborative efforts between researchers, engineers, and construction professionals will be crucial to unlock the full potential of VLPC and to overcome the existing limitations.

Author Contributions: Methodology, X.Y. and Q.C.; validation, Q.C. and J.Z.; formal analysis, Q.C.; investigation, Q.C.; resources, X.Y. and W.X.; data curation, J.Z.; writing—original draft preparation, J.Z.; writing—review and editing, W.X. and X.Y.; project administration, Q.C.; funding acquisition, Q.C. and W.X. All authors have read and agreed to the published version of the manuscript.

Funding: The authors would like to thank the National Natural Science Foundation of China (Grant No.: 51978678, Grant No.: 52178208), the National Science Fund for Distinguished Young Scholars of Beijing (Grant No.: JQ22026), the Natural Science Foundation of Guangdong, China (Grant No.: 2021A1515010613); the Science and Technology Program of Guangzhou, China (Grant No.: 202201010738), the Fundamental Research Funds for the Central Universities (Grant No.: 2023ZYGXZR031), and the Shenzhen Natural Science Foundation (Grant No.: JCYJ20230807114401002).

Data Availability Statement: Data are contained within the article.

Conflicts of Interest: The authors declare no conflicts of interest.

References

1. Wang, F.; Sun, C.; Ding, X.; Kang, T.; Nie, X. Experimental Study on the Vegetation Growing Recycled Concrete and Synergistic Effect with Plant Roots. *Materials* **2019**, *12*, 1855. [CrossRef] [PubMed]
2. Ganapathy, G.P.; Alagu, A.; Ramachandran, S.; Panneerselvam, A.S.; Vimal Arokiaraj, G.G.; Panneerselvam, M.; Panneerselvam, B.; Sivakumar, V.; Bidorn, B. Effects of fly ash and silica fume on alkalinity, strength and planting characteristics of vegetation porous concrete. *J. Mater. Res. Technol.* **2023**, *24*, 5347–5360. [CrossRef]
3. Sheng, Y.P.; Li, H.B.; Guan, B.W. Mix Design of Compaction-Free Porous Concrete Permeable Base. *Adv. Mater. Res.* **2012**, *368–373*, 1416–1419. [CrossRef]
4. Qu, H.; Wang, C.; Huang, X.; Ding, Y.; Huang, X. Seismic Performance of Substrate for Vegetation Concrete from Large-Scale Shaking Table Test. *Shock Vib.* **2020**, *2020*, 6670726. [CrossRef]
5. Shen, W.; Liu, Y.; Wu, M.; Zhang, D.; Du, X.; Zhao, D.; Xu, G.; Zhang, B.; Xiong, X. Ecological carbonated steel slag pervious concrete prepared as a key material of sponge city. *J. Clean. Prod.* **2020**, *256*, 120244. [CrossRef]
6. Ge, P.; Huang, W.; Zhang, J.; Quan, W.; Guo, Y. Mix proportion design method of recycled brick aggregate concrete based on aggregate skeleton theory. *Constr. Build. Mater.* **2021**, *304*, 124584. [CrossRef]
7. Endawati, J.; Rochaeti, R.; Utami, R. Optimization of Concrete Porous Mix Using Slag as Substitute Material for Cement and Aggregates. *Appl. Mech. Mater.* **2017**, *865*, 282–288. [CrossRef]
8. Bao, X.; Liao, W.; Dong, Z.; Wang, S.; Tang, W. Development of Vegetation-Pervious Concrete in Grid Beam System for Soil Slope Protection. *Materials* **2017**, *10*, 96. [CrossRef]
9. Lee, K.-H.; Yang, K.-H. Development of a neutral cementitious material to promote vegetation concrete. *Constr. Build. Mater.* **2016**, *127*, 442–449. [CrossRef]
10. Kim, H.-H.; Park, C.-G. Performance Evaluation and Field Application of Porous Vegetation Concrete Made with By-Product Materials for Ecological Restoration Projects. *Sustainability* **2016**, *8*, 294. [CrossRef]
11. Kim, H.-H.; Lee, S.-K.; Park, C.-G. Carbon Dioxide Emission Evaluation of Porous Vegetation Concrete Blocks for Ecological Restoration Projects. *Sustainability* **2017**, *9*, 318. [CrossRef]
12. Xu, J.H.; Chen, S.L.; Wang, Y.; Li, G. Research on Mixture Design and Mechanical Performance of Porous Asphalt Concrete in Cold Regions. *Adv. Mater. Res.* **2011**, *299–300*, 770–773. [CrossRef]
13. Cheng, J.; Luo, X.; Shen, Z.; Guo, X. Study on Vegetation Concrete Technology for Slope Protection and Greening Engineering. *Pol. J. Environ. Stud.* **2020**, *29*, 4017–4028. [CrossRef] [PubMed]
14. Kug, J.Y. A Study on the pH Reduction of Cement Concrete with Various Mixing Conditions. *J. Korea Inst. Build. Constr.* **2008**, *8*, 79–85.
15. Zhao, S.; Zhang, D.; Li, Y.; Gao, H.; Meng, X. Physical and Mechanical Properties of Novel Porous Ecological Concrete Based on Magnesium Phosphate Cement. *Materials* **2022**, *15*, 7521. [CrossRef]
16. Rattanashotinunt, C.; Tangchirapat, W.; Jaturapitakkul, C.; Cheewaket, T.; Chindaprasirt, P. Investigation on the strength, chloride migration, and water permeability of eco-friendly concretes from industrial by-product materials. *J. Clean. Prod.* **2018**, *172*, 1691–1698. [CrossRef]
17. Kong, J.; Cong, G.; Ni, S.; Sun, J.; Guo, C.; Chen, M.; Quan, H. Recycling of waste oyster shell and recycled aggregate in the porous ecological concrete used for artificial reefs. *Constr. Build. Mater.* **2022**, *323*, 126447. [CrossRef]
18. Su, R.; Qiao, H.; Li, Q.; Su, L. Study on the performance of vegetation concrete prepared based on different cements. *Constr. Build. Mater.* **2023**, *409*, 133793. [CrossRef]
19. Li, S.; Yin, J.; Zhang, G. Experimental investigation on optimization of vegetation performance of porous sea sand concrete mixtures by pH adjustment. *Constr. Build. Mater.* **2020**, *249*, 118775. [CrossRef]
20. Peng, H.; Yin, J.; Song, W. Mechanical and Hydraulic Behaviors of Eco-Friendly Pervious Concrete Incorporating Fly Ash and Blast Furnace Slag. *Appl. Sci.* **2018**, *8*, 859. [CrossRef]
21. Yang, Y.; Chen, J.; Zhou, T.; Liu, D.; Yang, Q.; Xiao, H.; Liu, D.; Chen, J.; Xia, Z.; Xu, W. Effects of freeze-thaw cycling on the engineering properties of vegetation concrete. *J. Environ. Manag.* **2023**, *345*, 118810. [CrossRef] [PubMed]
22. Kim, H.-H.; Park, C.-G. Plant Growth and Water Purification of Porous Vegetation Concrete Formed of Blast Furnace Slag, Natural Jute Fiber and Styrene Butadiene Latex. *Sustainability* **2016**, *8*, 386. [CrossRef]
23. *GB 175-2007*; General-Purpose Silicate Cement. China Standard Publishing House: Beijing, China, 2007.
24. *GB/T 20472-2006*; Sulphate Aluminate Cement. Beijing University of Technology Press: Beijing, China, 2006.
25. *GJJ/T 253-2016*; Technical Specification for Application of Pervious Recycled Aggregate Concrete. Industry Standards of the People's Republic of China: Beijing, China, 2009.
26. *CJJ/T 135-2009*; Technical Specifications for Pervious Concrete Pavement. Industry Standards of the People's Republic of China: Beijing, China, 2009.
27. *GB/T 50081-2019*; Standard for Mechanical Properties of Concrete Test Methods. China Standard Press: Beijing, China, 2019.
28. Loh, P.Y.; Shafigh, P.; Katman, H.Y.B.; Ibrahim, Z.; Yousuf, S. pH Measurement of Cement-Based Materials: The Effect of Particle Size. *Appl. Sci.* **2021**, *11*, 8000. [CrossRef]
29. Doussang, L.; Samson, G.; Deby, F.; Huet, B.; Guillon, E.; Cyr, M. Durability parameters of three low-carbon concretes (low clinker, alkali-activated slag and supersulfated cement). *Constr. Build. Mater.* **2023**, *407*, 133511. [CrossRef]
30. Li, L.; Nam, J.; Hartt, W.H. Ex situ leaching measurement of concrete alkalinity. *Cem. Concr. Res.* **2005**, *35*, 277–283. [CrossRef]

31. Al-Amoudi, O.S.B.; Maslehuddin, M.; Shameem, M.; Ibrahim, M. Shrinkage of plain and silica fume cement concrete under hot weather. *Cem. Concr. Compos.* **2007**, *29*, 690–699. [CrossRef]
32. Fu, Q.; Zhang, Z.; Wang, Z.; He, J.; Niu, D. Erosion behavior of ions in lining concrete incorporating fly ash and silica fume under the combined action of load and flowing groundwater containing composite salt. *Case Stud. Constr. Mater.* **2022**, *17*, e01659. [CrossRef]
33. Seifan, M.; Mendoza, S.; Berenjian, A. Mechanical properties and durability performance of fly ash based mortar containing nano- and micro-silica additives. *Constr. Build. Mater.* **2020**, *252*, 119121. [CrossRef]
34. Nochaiya, T.; Suriwong, T.; Julphunthong, P. Acidic corrosion-abrasion resistance of concrete containing fly ash and silica fume for use as concrete floors in pig farm. *Case Stud. Constr. Mater.* **2022**, *16*, e01010. [CrossRef]
35. Ibrahim, K.I.M. Recycled waste glass powder as a partial replacement of cement in concrete containing silica fume and fly ash. *Case Stud. Constr. Mater.* **2021**, *15*, e00630. [CrossRef]
36. Feng, W.; Tang, Y.; Zhang, Y.; Qi, C.; Ma, L.; Li, L. Partially fly ash and nano-silica incorporated recycled coarse aggregate based concrete: Constitutive model and enhancement mechanism. *J. Mater. Res. Technol.-JMRT* **2022**, *17*, 192–210. [CrossRef]
37. Ma, X.; He, T.; Xu, Y.; Yang, R.; Sun, Y. Hydration reaction and compressive strength of small amount of silica fume on cement-fly ash matrix. *Case Stud. Constr. Mater.* **2022**, *16*, e00989. [CrossRef]
38. Natkunarajah, K.; Masilamani, K.; Maheswaran, S.; Lothenbach, B.; Amarasinghe, D.A.S.; Attygalle, D. Analysis of the trend of pH changes of concrete pore solution during the hydration by various analytical methods. *Cem. Concr. Res.* **2022**, *156*, 106780. [CrossRef]
39. Kim, J.-S.; Kwon, S.; Choi, J.-W.; Cho, G.-C. Properties of low-PH cement grout as a sealing material for the geological disposal of radioactive waste. *Nucl. Eng. Technol.* **2011**, *43*, 459–468. [CrossRef]

Disclaimer/Publisher's Note: The statements, opinions and data contained in all publications are solely those of the individual author(s) and contributor(s) and not of MDPI and/or the editor(s). MDPI and/or the editor(s) disclaim responsibility for any injury to people or property resulting from any ideas, methods, instructions or products referred to in the content.

Article

Study on the Influence of Magnesite Tailings on the Expansion and Mechanical Properties of Mortar

Feifei Jiang [1,2], Juan Zhou [3,*], Zhongyang Mao [4,*] and Bi Chen [4]

1. School of Civil Engineering, Nantong Institute of Technology, Nantong 226000, China; 999620140019@just.edu.cn
2. Suzhou Institute of Technology, Jiangsu University of Science and Technology, Zhangjiagang 215600, China
3. School of Business, Nantong Institute of Technology, Nantong 226000, China
4. College of Materials, Science and Engineering, Nanjing Tech University, Nanjing 211800, China; bichen1900@126.com
* Correspondence: zhoujuan198808@163.com (J.Z.); mzy@njtech.edu.cn (Z.M.)

Abstract: To reduce the mining of high-grade magnesite and solve the environmental pollution caused by magnesite tailings, magnesite tailings were used to produce MgO expansion agent (MEA), and a detailed study of its performance was carried out in this study. Firstly, the effects of different calcination times on the calcination products, the specific surface area, and the activity of MEA were analyzed. Then, the MEA produced by calcinating at 950 °C for 1 h was taken as the research object, and the effects of its content on the expansion performance, compressive strength, and flexural strength of the mortar were studied. The results showed that the decomposition of magnesite tailings after high-temperature calcination produced MEA, and the longer the calcination time, the lower the activity. The calcined tailings could compensate for the shrinkage of the mortar, and the expansion increased with the increase in curing temperature. What is more, when the content was less than 8%, the hydration of MEA filled the pores and improved the compactness, so the strength of the mortar increased with the increase in the expansion agent content. When the dosage was greater than 8%, excessive expansion increased the porosity, causing harmful expansion of the mortar and damaging its integrity, leading to a decrease in strength. Fly ash reduced the expansion of mortar, and after adding 30% fly ash, the expansion decreased by 20.0–36.1%, and the ability to suppress expansion decreased with the increase in curing temperature.

Keywords: magnesite tailings; MgO expansion agent; mortar; compressive strength; flexural strength; porosity

1. Introduction

Cement has been widely used as the raw material for concrete. However, when cement comes into contact with water and reacts, its volume gradually decreases. When shrinkage is constrained, shrinkage tensile stress is generated, leading to structural cracking [1–3]. This not only reduces the strength of the concrete but also the bearing capacity of the structure and the durability of the concrete. but also makes it easier for harmful ions to enter the interior of the concrete, greatly reducing its service life, resulting in a huge loss of natural resources and economy. The problem of cracking caused by the shrinkage of cement-based materials has been puzzling engineers and researchers. And as the strength of concrete increases, the fineness of cement gradually decreases, the C_3S content increases, and the water-to-cement ratio gradually decreases. Although these changes improve the strength of concrete, the increased shrinkage of concrete makes it more prone to shrinkage cracks [4–6].

Adding an expansion agent to compensate for the shrinkage of cement by utilizing the expansion generated by its hydration is a widely recognized method [7–9]. Currently, the expansion agents commonly used in engineering are sulfoaluminate-type, MgO-type, and

CaO-type expansion agents. Different from the other two kinds of expansion agents, the MgO expansion agent (MEA) has many advantages, such as a lower water requirement for hydration, a more stable reaction-generated substance, $Mg(OH)_2$, an artificially adjustable expansion rate, and so on. So MEA is widely used in modern concrete [10–12]. Since the 1970s, a large number of Chinese researchers have begun to study MEA, which is applied to large-scale buildings such as dams and airport pavement. The shrinkage of this mass concrete is well compensated by the use of MEA with different activities during dry shrinkage and temperature shrinkage, which greatly reduces the shrinkage cracks and improves the durability of the mass concrete [13,14].

Nowadays, the MEA sold on the market is basically produced by calcining high-grade magnesite. However, with the continuous exploitation of non-renewable minerals, environmental pollution and resource depletion issues are becoming increasingly serious. Coupled with the government's restrictions on the mining of non-renewable mineral resources, the grade of magnesite resources is decreasing [15–19]. At the same time, a large amount of low-grade magnesite is discarded every year. These randomly discarded tailings have caused serious resource waste and also occupied a large amount of land. Therefore, how to reduce the mining of high-quality magnesite and solve the environmental pollution caused by tailings has become a very urgent problem.

Some scholars have attempted to use other methods to produce MEA. Temiz [20] prepared an expansion agent containing MgO using calcined dolomite ($CaMg(CO_3)_2$). However, this expansion agent also contained a large amount of CaO, which had a fast hydration rate and generated a large amount of expansion before the concrete hardened. This expansion could not compensate for the shrinkage of hardened concrete. Gao [21] used industrial by-products (magnesite and serpentine) to produce MgO expansion agents. It was found that the products of industrial by-products after calcination could also produce expansion, which could compensate for the shrinkage of dam concrete. However, compared to the concrete used in buildings and bridges, the strength of dam concrete is not high, and it is not clear whether the expansion agent produced by by-products has the same effect on shrinkage compensation as high-strength concrete (greater than 50 MPa). There are still many concerns about whether the expansion agent produced by by-products will reduce the strength of high-strength concrete. The above research confirms the necessity and feasibility of using other methods to produce expansion agents. But there is still a lack of detailed research on the effects of calcination temperature and time on the mineral composition, specific surface area, and activity of MEA, which restricts the development of industrial by-products to produce expansion agents. The above issues are also likely the main reason for the increasingly scarce and expensive prices of magnesite.

This paper used magnesite tailings from Laizhou (China) to produce MEA, studied the influence of different calcination methods on the performance of expansion agents, and analyzed the influence of tailings on the strength and expansion performance of mortar, which provided a new way to solve the environmental pollution of magnesite tailings. And at the same time, the study can effectively reduce the price of MEA and promote the promotion and utilization of expansion agents.

2. Materials and Methods

2.1. Materials

2.1.1. Magnesite Tailing

The magnesite tailings were from Laizhou (Shandong, China), and their chemical composition is shown in Table 1. It can be seen from Table 1 that the content of MgO is lower than 43%, the content of SiO_2 is higher than 3.5%, and the content of CaO is higher than 1.5%. The main impurities in tailings are chlorite (($Mg,Fe)_{4.75}Al_{1.25}[Al_{1.25}Si_{2.75}O_{10}](OH)_8$) and talc ($Mg_3[Si_4O_{10}](OH)_2$), as well as a small amount of dolomite and calcite. According to the classification of magnesite (Table 2), the useful component MgO in tailings is insufficient, while other impurities exceed the standard, making it not one of the four types of ores. It is a low-value tailing with no benefits for mining or processing, so it was often discarded arbitrarily in the past, polluting the environment. The mineral composition of magnesite tailings is shown in Figure 1.

Table 1. Chemical compositions of magnesite tailing.

Material	Chemical Compositions (%)						
	CaO	MgO	Al_2O_3	SiO_2	Fe_2O_3	Loss	Total
Magnesite tailing	3.91	41.67	2.45	5.67	0.93	45.13	99.76

Table 2. Magnesite classification.

Grade	Chemical Compositions (%)		
	MgO	CaO	SiO_2
Top Grade	≥47	≤0.6	≤0.6
Grade 1	≥46	≤0.8	≤1.2
Grade 2	≥45	≤1.5	≤1.5
Grade 3	≥43	≤1.5	≤3.5

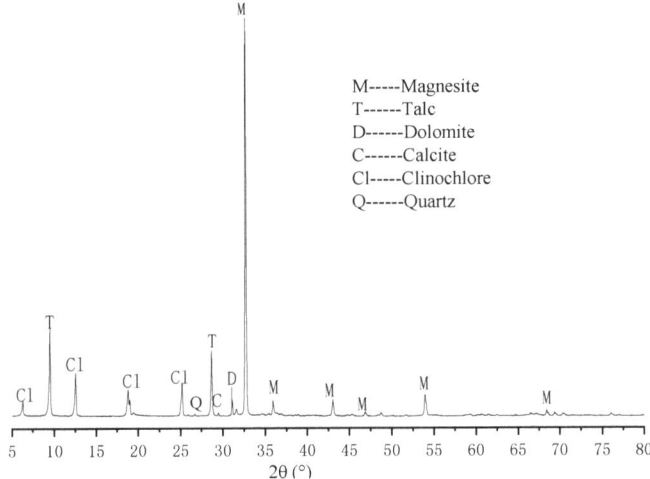

Figure 1. XRD pattern of magnesite tailing.

2.1.2. Cement

The cement was PII 52.5 Portland cement produced by Jiangnan Xiaoyetian Cement Co., Ltd. (Nanjing, China). Its chemical composition is shown in Table 3, and its mineral composition is shown in Figure 2. It can be seen from Figure 2 that the main mineral composition of cement is C_3S, C_2S, C_3A, and C_4AF, with a small amount of gypsum and calcite. The particle size distribution of cement is shown in Figure 3.

Table 3. Chemical compositions of cement.

Material	Chemical Compositions (%)							
	CaO	MgO	Al_2O_3	SiO_2	Fe_2O_3	SO_3	Loss	Total
Cement	64.73	0.89	4.39	19.41	2.97	2.59	2.40	97.38

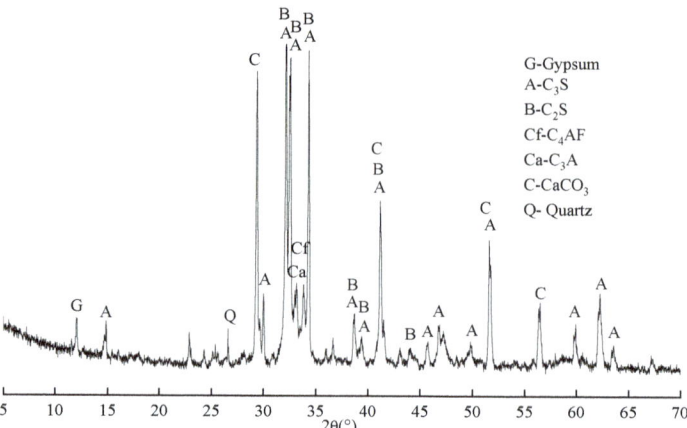

Figure 2. XRD pattern of Portland cement (P·II52.5).

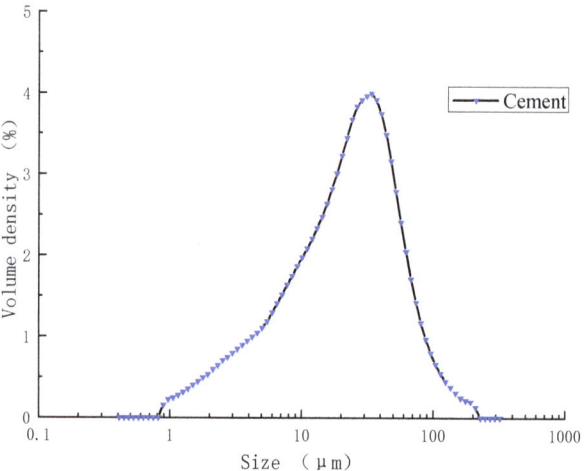

Figure 3. Particle size distribution of cement.

2.1.3. Fly Ash

Fly ash was class I fly ash from Huaneng Power Plant (Nanjing, China). Chemical composition is shown in Table 4, and mineral composition is shown in Figure 4. The particle size distribution of fly ash is shown in Figure 5.

Table 4. Chemical compositions of fly ash.

Material	Chemical Compositions (%)						
	MgO	CaO	SiO_2	Al_2O_3	Fe_2O_3	Loss	Total
Fly ash	0.92	4.40	47.84	34.71	7.31	3.02	98.20

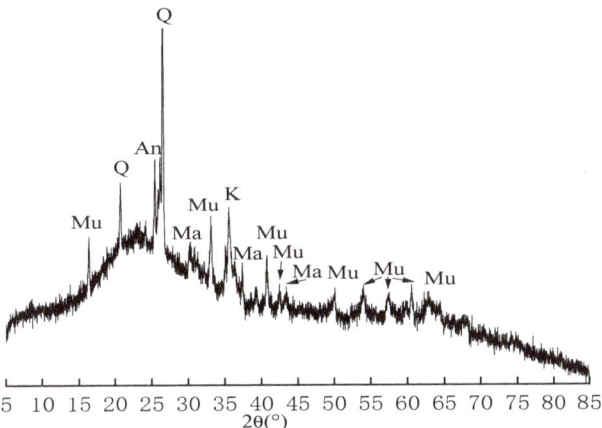

Figure 4. XRD pattern of fly ash: An—Andlusite; Mu—Mullite; Q—Quartz; Ma—Magnetite; K—Kyanite.

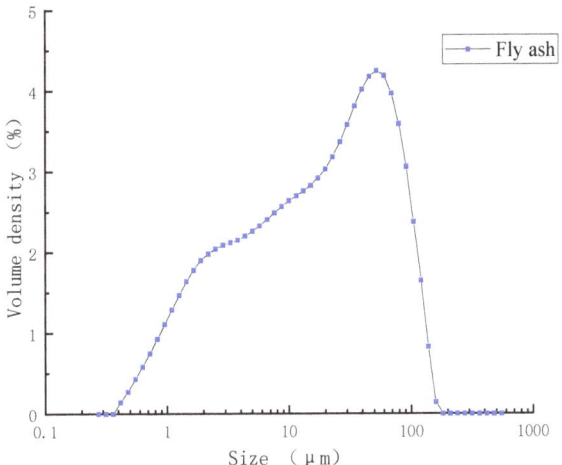

Figure 5. Particle size distribution of fly ash.

2.1.4. Sand

The fine aggregate was selected from natural river sand from Nantong (Jiangsu, China). It was tested according to the Chinese standard GB/T 14684-2022 [22], and the test results are shown in Table 5 and Figure 6. The fineness modulus of river sand was 2.1, which belonged to medium sand. The measured cumulative sieve residue of river sand is between the specified upper and lower limits, meeting the grading requirements.

Table 5. The properties of river sand.

Apparent Density (g/cm^3)	Bulk Density (g/cm^3)	Porosity (%)
2.63	1.51	43.5

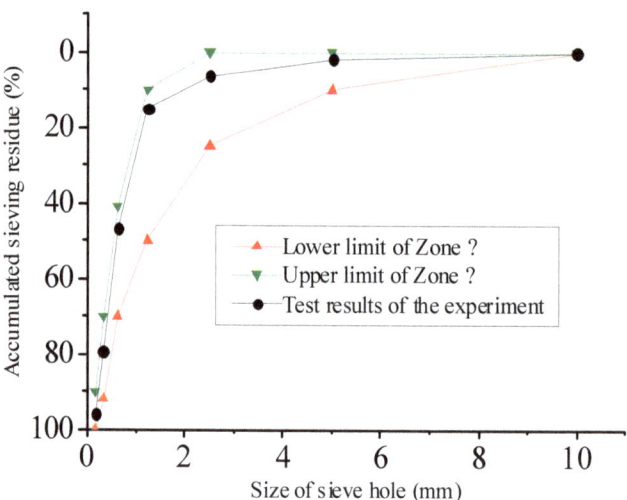

Figure 6. Grading curve of river sand.

2.1.5. Mix Design

The sand-to-binder ratio (S/B) and water-to-binder ratio (W/B) of mortar were fixed at 3 and 0.5, respectively. Four kinds of mortars with different MEA contents were prepared, and the replacement rates of MEA to cement were 4%, 8%, 12%, and 16%, respectively. In order to study the effect of fly ash on tailing hydration, two different proportions of fly ash (0% and 30%) were used instead of cement.

2.2. Experimental Works

2.2.1. Calcination of Tailings

The magnesite tailings were crushed by large, medium, and small jaw crushers in turn. Then they were milled in a mill for 1.0 h and then screened by a 0.016 mm square hole sieve. The method of calcination after briquetting was adopted in the test. After stirring 50 g of sieved tailings with 2.0% water and holding the pressure at 6 MPa for 5 s, a cube of 5 cm × 5 cm × 1 cm was made. Then it was calcined in a box furnace. The calcination temperature was 950 °C, the calcination time was 0.5 h, 1.0 h, 1.5 h, and 2.0 h, respectively, and the heating rate was 6 °C/min. The calcined tailing was cooled rapidly in the air and passed through a square hole sieve of 0.08 mm after being milled for 3 S by a vibration mill.

2.2.2. Performance Test of MEA

(1) Mineral composition

The mineral composition of MEA was analyzed by X-ray diffraction produced in the United States, model ARL XTRA.

(2) Specific surface area of MEA

The specific surface area test of MEA was performed by a specific surface area and micropore analyzer, which is produced in the United States and is model ASAP2020M.

(3) Measurement of f-CaO content

The content of f-CaO in MEA was tested according to the glycerin alcohol substitute method in "Cement Chemical Analysis Method" (GB/T 176-2008 [23]).

(4) Measurement of MgO content

The content of MgO in MEA was tested by X-ray diffraction-phase quantitative analysis (internal standard k-value method). The internal standard material was ZnO, the content was 10%, the scanning range was 35–45°, and the scanning speed was 1 °/min.

2.2.3. Activity of MEA

First, a beaker containing 200 mL of distilled water and 2 drops of phenolphthalein indicator was placed in a constant-temperature water bath at 30 °C. Then, 1.7 g of MEA and some citric acid were placed in a beaker and timed with a stopwatch until the solution turned red. The time recorded by the stopwatch was the active value of MEA. The amount of citric acid is calculated by Formula (1):

$$M = 2.84 \times X + 2.13 \times Y \tag{1}$$

where M is the amount of citric acid; X is the content of MgO in MEA; and Y is the content of f-CaO in MEA.

2.2.4. Mortar Test

(1) Expansion test

The mortar specimen was a cuboid of 40 mm × 40 mm × 160 mm. The effect of MEA on expansion performance was studied by measuring the length variation in mortar specimens. The calcination temperature of MEA was 950 °C, and the calcination time was 1 h. The specimen was demolished after standard curing for 24 h after pouring, and its initial length (L_0) was measured after curing in 20 °C water for 2 h. Then the test specimens were put into water at 20 °C for long-term curing, and the length (L_1) was measured after the corresponding age. Three identical specimens were made for each mix proportion of mortar, and the average expansion of the three specimens was considered the expansion of the mortar. The calculation formula for the expansion rate (φ) of the test piece was:

$$\varphi = (L_1 - L_0)/L \tag{2}$$

where L is the initial effective length of the mortar specimen, taken at 150 mm.

(2) Strength test

The strength test was carried out in accordance with the "Test Method for Strength of Cement Mortar" (GBT 17671-1999 [24], China), and the mix proportion was the same as that of the expansion test. Three identical samples were prepared for each dosage in the flexural strength test. Six identical samples were prepared for each dosage in the compressive strength test. The average value of the sample was considered the final strength.

(3) Morphology test

The morphology of mortar containing MEA was examined by SEM (HR-8100, Hitachi, Ltd., Tokyo, Japan).

(4) Porosity and pore size distribution

The porosity and pore size distribution of mortar were measured by a GT-60 mercury intrusion tester produced by Canta Instruments Co., Ltd. (Florida, USA).

3. Results

3.1. The Effect of Calcination Time on the Calcined Product of Magnesite

Figure 7 shows the mineral composition analysis results of MEA produced by calcining magnesite tailings at 950 °C for 0.5 h, 1.0 h, 1.5 h, and 2.0 h, respectively. From the comparison of minerals before and after calcination in Figures 1 and 7, it can be seen that when the calcination temperature and calcination time were up to 950 °C for 1 h, no magnesite, chlorite, talc, calcite, or dolomite were found in the XRD pattern. This means that there is only a small amount left that cannot be displayed on the XRD pattern.

In addition, the XRD pattern shows that the main minerals obtained after 0.5 h of heat preservation were periclase (MgO), as well as a small amount of quartz (SiO_2), lime (CaO), and lamite (C_2S). With the prolongation of calcination time, the peaks of MgO, SiO_2, and C_2S changed slightly. When the calcination time increased from 0.5 h to 2.0 h, the peak height of MgO increased slightly, the peak height of SiO_2 and CaO decreased gradually, and the peak height of C_2S increased obviously.

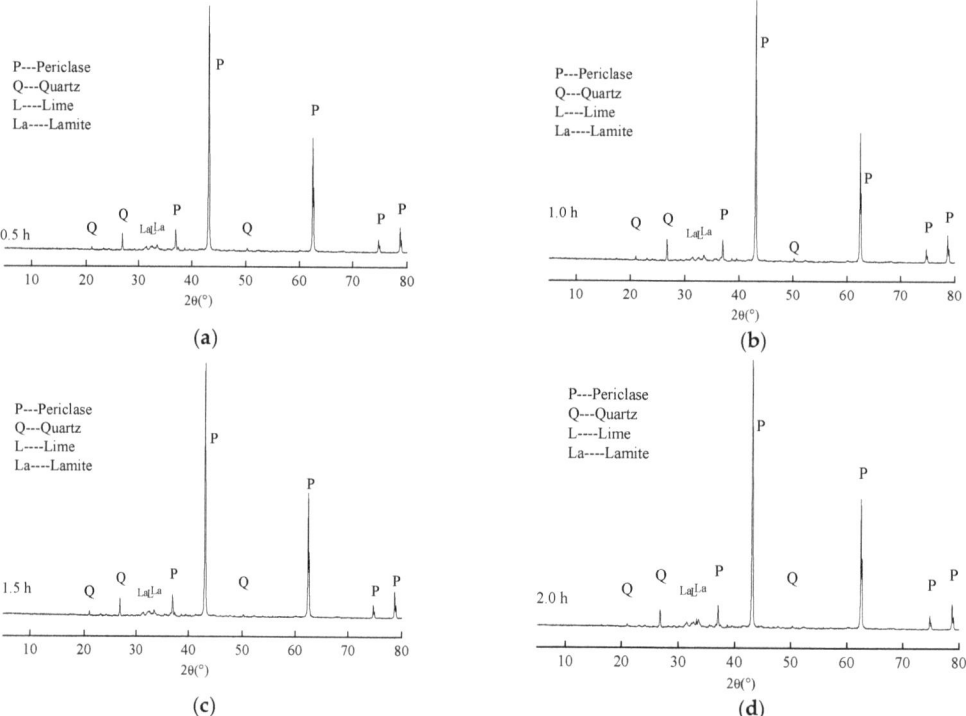

Figure 7. XRD patterns of MEA calcining magnesite tailings at 950 °C for different times: (**a**) 0.5 h, (**b**) 1.0 h, (**c**) 1.5 h, and (**d**) 2.0 h.

Table 6 lists the content of MgO in MEA produced by calcining magnesite tailings at 950 °C for 0.5 h, 1.0 h, 1.5 h, and 2.0 h. It can be seen from Table 6 that the content of MgO increased slightly with the prolongation of the calcination time, which indicates that when the calcination time was 0.5 h, there were still a small amount of magnesia minerals that were not completely decomposed. With the prolongation of the calcination time, the decomposition of magnesia minerals was more sufficient, increasing the content of MgO. According to Figure 7 and Table 6, when the calcination time was more than 1.5 h, $MgCO_3$, calcite, dolomite, and other minerals in the tailings had almost completely decomposed. Even if the calcination time continued to increase, the content of MgO in MEA would not change greatly.

Table 6. Contents of MgO in MEA at 950 °C for 0.5 h, 1.0 h, 1.5 h and 2.0 h.

Material	Content of f-MgO (%)			
	0.5 h	1.0 h	1.5 h	2.0 h
MEA	73.33	73.48	73.60	73.68

Table 7 lists the content of f-CaO in MEA produced from calcined magnesite tailings at 950 °C for 0.5 h, 1.0 h, 1.5 h, and 2.0 h. It can be seen from Table 7 that the content of f-CaO decreased with the prolongation of calcination time. This phenomenon, combined with the smaller diffraction peak of SiO_2 in Figure 7, collectively indicates that as the calcination time increased, f-CaO reacted with SiO_2 to form C_2S in a solid state [18].

Table 7. Contents of f-CaO in MEA at 950 °C for 0.5 h, 1.0 h, 1.5 h, and 2.0 h.

Material	Content of f-CaO (%)			
	0.5 h	1.0 h	1.5 h	2.0 h
MEA	0.94	0.88	0.82	0.56

3.2. Effect of Holding Time on Specific Surface Area of MEA

Table 8 shows the specific surface area of MEA produced by calcined magnesite tailings at 950 °C for 0.5 h, 1.0 h, 1.5 h, and 2.0 h. It can be seen from Table 8 that the specific surface area of MEA decreased gradually with the extension of the calcination time. This is because the longer the calcination time, the better the crystallization of MEA and the larger the crystal size, resulting in a smaller specific surface area of MEA.

Table 8. Specific surface area of MEA at 950 °C for 0.5 h, 1.0 h, 1.5 h, and 2.0 h.

Material	Specific Surface Area ($m^2 \cdot g^{-1}$)			
	0.5 h	1.0 h	1.5 h	2.0 h
MEA	22.34	20.36	19.45	18.76

3.3. Effect of Calcination Time on MEA Activity

Table 9 lists the hydration activities of MEA from calcined magnesite tailings with 0.5 h, 1.0 h, 1.5 h, and 2.0 h holding at 950 °C. From Table 9, it can be seen that the activity of MEA decreased with increasing calcination time. This is because the longer the calcination time, the fewer the crystal defects and the smaller the specific surface area of MEA, resulting in a slower reaction rate and thus prolonging the reaction time. Therefore, we can produce different active MEA by changing the calcination time and obtain different expansions to meet the needs of different constructions.

Table 9. Hydration activity of MEA at 950 °C for 0.5 h, 1.0 h, 1.5 h, and 2.0 h.

Material	Hydration Activity (s)			
	0.5 h	1.0 h	1.5 h	2.0 h
MEA	104	153	187	203

3.4. Effect of MEA on Mortar Deformation

Figure 8 shows the effect of MEA on the expansion of cement mortar when curing in water at 20 °C and 30 °C. From Figure 8, it can be observed that the expansion rate of MEA mortar was fast in the early stage (within 30 d). Subsequently, as MEA was continuously consumed, its expansion gradually decreased, and there was no phenomenon of expansion regression, indicating that the products after MEA hydration were stable. The increase in curing temperature could promote the hydration of MEA and increase the expansion of mortar. This promoting effect was more obvious in the early stage (within 30 d). For mortar with 8% MEA added, increasing the curing temperature resulted in an increase of 58.3% and 25.0% in expansion after 3 d and 28 d, respectively. This means that during actual construction, we can increase the early curing temperature to improve the ability of MEA to compensate for shrinkage.

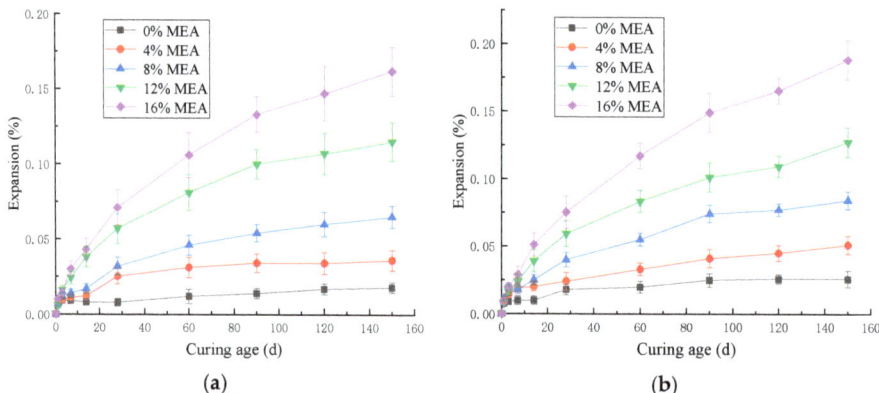

Figure 8. The effect of MEA on expansion of mortars cured in water: (**a**) 20 °C; (**b**) 30 °C.

Figure 9 shows the effect of MEA on the expansion of mortar mixed with 30% fly ash. It can be seen from Figure 9 that with the increase in MEA content, the expansion of mortar mixed with 30% fly ash cured in 20 °C or 30 °C water increased gradually. Compared with Figures 8 and 9, the expansion rate of MEA mortar was reduced by adding fly ash. Under the condition of water curing at 20 °C, after adding fly ash, the expansion of mortar mixed with 8%, 12%, and 16% MEA decreased by 35.0%, 30.8%, and 36.1%, respectively, at 120 d. The main reason for this change is that $Mg(OH)_2$ reacted with SiO_2 in fly ash to form C-S-H gel, which consumed the expansion product of MEA hydration, and the generated gel covered the surface of MEA, inhibiting the continued hydration of MEA and thereby reducing the expansion of the mortar [7].

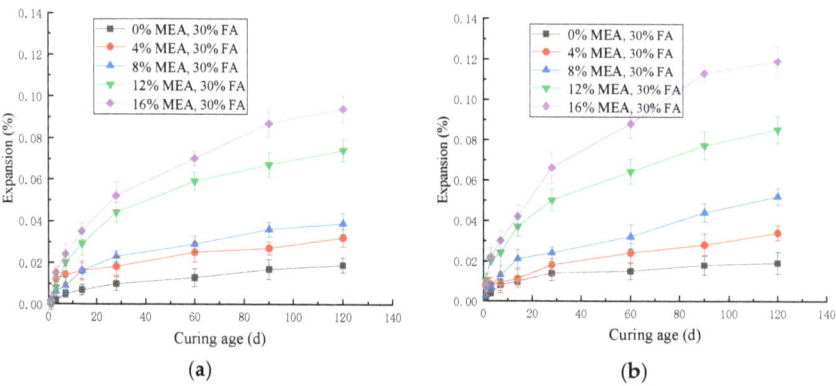

Figure 9. The effect of MEA on the expansion of mortars with 30% fly ash: (**a**) 20 °C; (**b**) 30 °C.

By comparing Figure 8b with Figure 9b, it can be found that after adding fly ash, the expansion of mortar mixed with 8%, 12%, and 16% MEA decreased by 32.5%, 22.0%, and 27.9%, respectively, at 30 °C for 120 d. It can be seen that fly ash inhibited the expansion of MEA, and the degree of inhibition decreased with the increase in curing temperature.

3.5. Influence of MEA on Mortar Strength

Figure 10 shows the influence of MEA on the compressive strength and flexural strength of mortar without fly ash. As can be seen from Figure 10, when the content of MEA was less than 4%, the strength of the mortar increased with the increase in the content of MEA. Compared with ordinary mortar, the compressive strength and flexural strength of 4% MEA mortar increased by 12.18% and 7.94%, respectively, at 28 d. However, when the content of MEA was greater than 8%, the compressive strength and flexural strength of the mortar decreased. And the larger the content of MEA was, the more the strength of the mortar decreased. The compressive strength and flexural strength of mortar with 12% MEA decreased by 7.97% and 9.99%, respectively, at 90 d, and mortar with 16% MEA decreased by 21.64% and 17.48%, respectively. Therefore, from the perspective of ensuring the strength of mortar, the content of MEA should not exceed 8%.

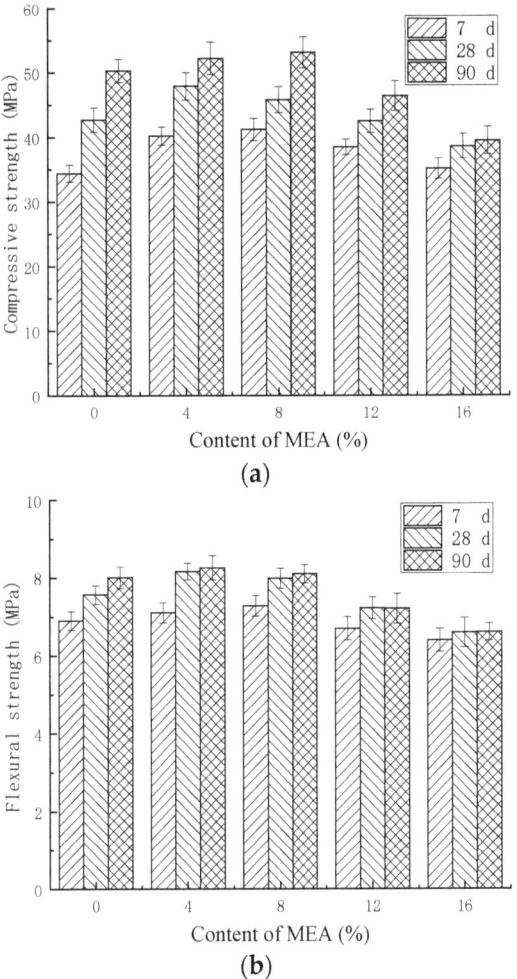

Figure 10. The strength of mortars with MEA: (**a**) compressive strength; (**b**) flexural strength.

Figure 11 shows the microscopic morphology of mortar without the addition of MEA. From Figure 11, it can be observed that when MEA was not added, due to the shrinkage of the mortar, the structure was loose and the compactness was poor. It can be clearly seen from the picture that there were many small holes. This is also one of the main reasons why the compressive strength and flexural strength of ordinary mortar were not high.

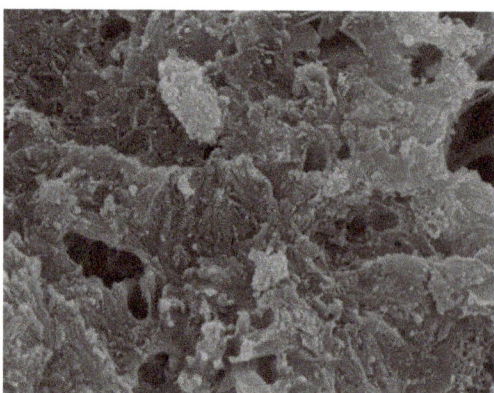

Figure 11. Mortar without MEA at 28 d.

Figures 12 and 13 show the SEM of mortar with 4% MEA and 16% MEA added at 28 d, respectively. From Figure 12, it can be observed that the calcined tailings decomposed to produce MgO. At 28 d, the hydration of MEA generated brucite, resulting in volume expansion. Brucite was in a columnar or sheet-like structure, interwoven together to fill the voids in the concrete, which explains why the strength of the mortar mixed with 4% MEA increased at 28 d. From Figure 13, it can be observed that when the content of MEA increased to 16%, MEA hydration generated a large amount of expansion, which damaged the integrity of the mortar and caused cracks, leading to a decrease in mortar strength.

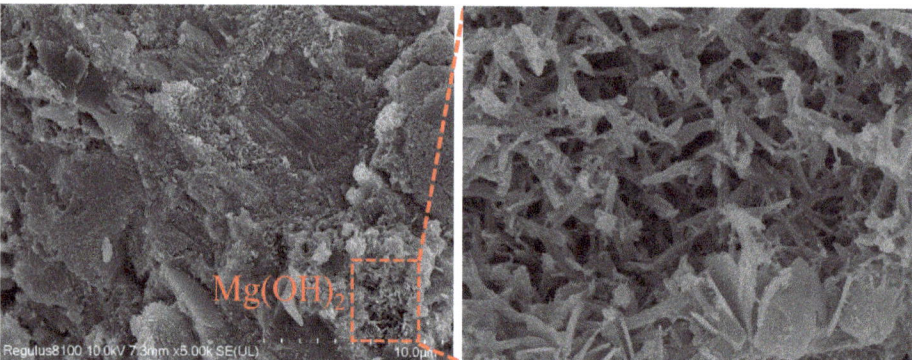

Figure 12. Mortar with 4% MEA at 28 d.

3.6. The Pore Size Distribution and Porosity of MEA for Mortar

According to ref. [25], the pores in concrete are divided into four categories based on their pore size: harmless pores (less than 20 nm), less harmful pores (20–100 nm), harmful pores (100–200 nm), and more harmful pores (greater than 200 nm). Figure 14 shows the effect of MEA on the pore size distribution of mortar. From Figure 14, it can be seen that when 4% expansion agent was added, as the curing time extended, the expansion agent hydrated and expanded to fill the voids, resulting in a decrease in the total porosity of the mortar, making it denser, especially for larger pores. The porosity at 28 d decreased by 19% compared to 7 d.

Figure 13. Mortar with 16% MEA at 28 d.

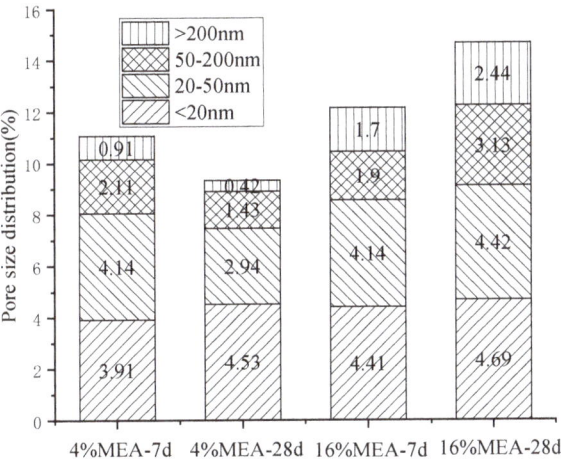

Figure 14. Pore size distribution of mortars containing MEA.

However, when the content of MEA increased to 16%, the porosity of the mortar increased by 30.4% at 7 d, and as the curing time increased, the porosity further increased, indicating that excessive expansion caused harmful expansion and damaged the integrity of the mortar. The results of porosity testing are consistent with those of strength testing. When the dosage was less than 8%, the expansion agent compensated for the shrinkage of the mortar, reduced the porosity, and optimized the pore structure. Therefore, the strength increased with the increase in MEA content. The reduction in porosity will also improve the impermeability of concrete, which will improve its durability [26–28]. On the other hand, when the dosage reached 16%, excessive MEA would lead to excessive expansion, resulting in more macropores at 28 d compared with pores at 7 d. Therefore, in future engineering applications, it is not only necessary to consider the calcination time of tailings but also to carefully select a reasonable content.

The freezing point of the pore water in the cementitious pore is very low (about −78 °C), which is not easy to freeze and has little influence on frost resistance. Although the freezing point of pore water in macroporous pore is high (about −1 °C), it is not easily saturated and has a buffering effect after freezing, so its impact on frost resistance is also small [29]. Among these three types of pores, the capillary pore has a significant impact on the frost resistance of concrete. It has a freezing point of −12 °C and is easily filled with water. From the pore structure data (Figure 14), after adding 4% MEA, the porosity of the capillary pore decreased at 28 d, indicating that the expansion agent can improve the frost resistance of concrete, which is of great significance for pouring large volumes of concrete in cold cities in Northern China. However, similarly, excessive MEA can also increase the porosity of the pores, so adding expansion agents reasonably is of great significance for improving the durability of concrete.

4. Conclusions

Considering the gradual decrease in high-grade magnesite and the environmental pollution caused by magnesite tailings, this paper attempted to use magnesite tailings to produce MEA. In this way, it can not only effectively protect the environment but also reduce the price of MEA and promote the promotion and application of MEA. This paper analyzed the effect of different calcination times on the material properties of MEA and focused on the effect of MEA produced at 950 °C for 1 h on the performance of mortar. The main conclusions are as follows:

(1) Similar to high-grade magnesite, magnesite tailings can also be used to produce MEA with different activities and different expansion properties by changing the calcination time. We can set reasonable production conditions to make use of magnesite tailings, which not only reduces environmental pollution but also compensates for the shrinkage of concrete. The longer the calcination time, the better the crystallization of MEA, the fewer crystal defects, and the smaller the specific surface area of MEA, resulting in a lower activity of MEA.

(2) The expansion of the mortar increased with the increase in MEA content. The early expansion rate was relatively high. With the consumption of MEA, the expansion in the later stage gradually slowed down and tended to stabilize, and there was no expansion regression in the later stage, indicating that the hydration products were stable. Therefore, using calcined magnesite tailings to compensate for concrete shrinkage is effective and safe.

(3) The expansion of MEA increased with the increase in curing temperature, and this increase was more pronounced in the early stage (within 30 d). At 7 d, the expansion of mortar cured at 30 °C increased by 58.3% compared to that cured at 20 °C. Therefore, we can increase the early curing temperature during construction to improve the ability of MEA to compensate for shrinkage.

(4) When the content did not exceed 8%, the strength of the mortar increased with the increase in MEA. In the early stage, the porosity of the mortar was large, and with the increase in curing time, the expansion agent hydration expanded to fill the pores. Compared to 7 d, the porosity of 28 d was reduced by 19%, and the pore structure was optimized, thus improving the strength and durability of the mortar. On the other hand, when the content was greater than 8%, the MEA produced harmful expansion, reducing the strength of the mortar. Especially when the content reached 16%, MEA produced excessive expansion, resulting in the enlargement of the porosity at 28 d, and this harmful expansion led to cracks in the mortar, destroying the integrity of the mortar. Therefore, it is recommended that the content of expansion agents not exceed 8% in engineering construction.

(5) After using MEA and fly ash simultaneously, fly ash inhibited the expansion of MEA. A total of 30% fly ash reduced the expansion of mortar by 20.0–36.1%, and the inhibitory ability decreased with the increase in curing temperature. In order to reveal the underlying reasons for the reduction in expansion caused by fly ash, future research should further analyze the impact of fly ash on the hydration of MEA. In addition, the impact of tailings on the durability of mortar and concrete also needs to be studied. Therefore, the influence of tailings on concrete properties will be discussed comprehensively, which provides a theoretical basis for engineering application.

Author Contributions: Methodology, F.J.; Formal analysis, F.J.; Investigation, J.Z. and B.C.; Resources, F.J.; Data curation, J.Z. and Z.M.; Writing—original draft, J.Z.; Writing—review and editing, F.J. and B.C.; Visualization, Z.M. All authors have read and agreed to the published version of the manuscript.

Funding: This research was funded by the Science and Technology Development Plan of Suzhou (SNG2022015, 2023SS14).

Institutional Review Board Statement: Not applicable.

Informed Consent Statement: Not applicable.

Data Availability Statement: All data generated or analyzed in this research were included in this published article. Additionally, readers can access all data used to support the conclusions of the current study from the corresponding author upon request.

Conflicts of Interest: The authors declare no conflict of interest.

References

1. Kouta, N.; Saliba, J.; Saiyouri, N. Effect of Flax Fibers on Early Age Shrinkage and Cracking of Earth Concrete. *Constr. Build. Mater.* **2020**, *254*, 119315. [CrossRef]
2. Ghourchian, S.; Wyrzykowski, M.; Plamondon, M.; Lura, P. On the Mechanism of Plastic Shrinkage Cracking in Fresh Cementitious Materials. *Cem. Concr. Res.* **2019**, *115*, 251–263. [CrossRef]
3. Ghourchian, S.; Wyrzykowski, M.; Baquerizo, L.; Lura, P. Susceptibility of Portland Cement and Blended Cement Concretes to Plastic Shrinkage Cracking. *Cem. Concr. Compos.* **2018**, *85*, 44–55. [CrossRef]
4. Şahmaran, M.; Li, V.C. Durability Properties of Micro-Cracked ECC Containing High Volumes Fly Ash. *Cem. Concr. Res.* **2009**, *39*, 1033–1043. [CrossRef]
5. Jacobsen, S.; Sellevold, E.J.; Matala, S. Frost Durability of High Strength Concrete: Effect of Internal Cracking on Ice Formation. *Cem. Concr. Res.* **1996**, *26*, 919–931. [CrossRef]
6. Shayanfar, M.A.; Farnia, S.M.H.; Ghanooni-Bagha, M.; Massoudi, M.S. The Effect of Crack Width on Chloride Threshold Reaching Time in Reinforced Concrete Members. *Asian J. Civ. Eng.* **2020**, *21*, 625–637. [CrossRef]
7. Rodríguez-Álvaro, R.; González-Fonteboa, B.; Seara-Paz, S.; Hossain, K.M.A. Internally Cured High Performance Concrete with Magnesium Based Expansive Agent Using Coal Bottom Ash Particles as Water Reservoirs. *Constr. Build. Mater.* **2020**, *251*, 118977. [CrossRef]
8. Liu, F.; Shen, S.L.; Hou, D.W.; Arulrajah, A.; Horpibulsuk, S. Enhancing Behavior of Large Volume Underground Concrete Structure Using Expansive Agents. *Constr. Build. Mater.* **2016**, *114*, 49–55. [CrossRef]
9. Li, M.; Liu, J.; Tian, Q.; Wang, Y.; Xu, W. Efficacy of Internal Curing Combined with Expansive Agent in Mitigating Shrinkage Deformation of Concrete Under Variable Temperature Condition. *Constr. Build. Mater.* **2017**, *145*, 354–360. [CrossRef]
10. Dung, N.; Unluer, C. Improving the Performance of Reactive MgO Cement-Based Concrete Mixes. *Constr. Build. Mater.* **2016**, *126*, 747–758. [CrossRef]
11. Kabir, H.; Hooton, R.D. Evaluating Soundness of Concrete Containing Shrinkage-Compensating MgO Admixtures. *Constr. Build. Mater.* **2020**, *253*, 119141. [CrossRef]
12. Nguyen, V.C.; Tong, F.G.; Nguyen, V.N. Modeling of Autogenous Volume Deformation Process of RCC Mixed with MgO Based On Concrete Expansion Experiment. *Constr. Build. Mater.* **2019**, *210*, 650–659. [CrossRef]
13. Huang, K.; Shi, X.; Zollinger, D.; Mirsayar, M.; Wang, A.; Mo, L. Use of MgO Expansion Agent to Compensate Concrete Shrinkage in Jointed Reinforced Concrete Pavement Under High-Altitude Environmental Conditions. *Constr. Build. Mater.* **2019**, *202*, 528–536. [CrossRef]
14. Mao, W.; Guo, S. Discovery and Significance of Quaternary Aqueously Deposited Aeolian Sandstones in the Sanhu Area, Qaidam Basin, China. *Petrol. Sci.* **2018**, *15*, 41–50. [CrossRef]
15. Yang, N.; Ning, P.; Li, K.; Wang, J. MgO-based Adsorbent Achieved from Magnesite for CO_2 Capture in Simulate Wet Flue Gas. *J. Taiwan Inst. Chem. E* **2018**, *86*, 73–80. [CrossRef]
16. Gehringer, S.; Luckeneder, C.; Hrach, F.; Flachberger, H. Processing of Caustic Calcined Magnesite (Magnesium Oxide) by the Use of Triboelectrostatic Belt Separation. *BHM Berg- Hüttenmännische Monatshefte* **2019**, *164*, 303–309. [CrossRef]
17. Knoll, C.; Müller, D.; Artner, W.; Welch, J.M.; Eitenberger, E.; Friedbacher, G.; Werner, A.; Weinberger, P.; Harasek, M. Magnesium Oxide from Natural Magnesite Samples as Thermochemical Energy Storage material. *Energy Procedia* **2019**, *158*, 4861–4869. [CrossRef]
18. Paholič, G.; Mateová, K. Stimulating the Thermal Decomposition of Magnesite. *Thermochim. Acta* **1996**, *277*, 75–84. [CrossRef]
19. Liu, Z.; Wang, S.; Huang, J.; Wei, Z.; Guan, B.; Fang, J. Experimental Investigation on the Properties and Microstructure of Magnesium Oxychloride Cement Prepared with Caustic Magnesite and Dolomite. *Constr. Build. Mater.* **2015**, *85*, 247–255. [CrossRef]
20. Temiz, H.; Kantarcı, F.; Inceer, M.E. Influence of Blast-Furnace Slag on Behaviour of Dolomite Used as a Raw Material of MgO-type Expansive Agent. *Constr. Build. Mater.* **2015**, *94*, 528–535. [CrossRef]

21. Gao, P.; Lu, X.; Geng, F.; Li, X.; Hou, J.; Lin, H.; Shi, N. Production of MgO-type Expansive Agent in Dam Concrete by Use of Industrial By-Products. *Build. Environ.* **2008**, *43*, 453–457. [CrossRef]
22. *GB/T 14684-2022*; Sand for Construction. National Standard of The People's Republic of China: Beijing, China, 2022. (In Chinese)
23. *GB/T 176-2008*; Methods for Chemical Analysis of Cement. National Standard of The People's Republic of China: Beijing, China, 2008. (In Chinese)
24. *GBT 17671-1999*; Method of Testing Cements—Determination of Strength. National Standard of The People's Republic of China: Beijing, China, 1999. (In Chinese)
25. Wu, Z.W.; Lian, H.Z. *High Performance Concrete*; China Railway Press: Beijing, China, 1999; pp. 22–27. (In Chinese)
26. Mohan, M.K.; Rahul, A.; Van Stappen, J.F.; Cnudde, V.; De Schutter, G.; Van Tittelboom, K. Assessment of pore structure characteristics and tortuosity of 3D printed concrete using mercury intrusion porosimetry and X-ray tomography. *Cem. Concr. Compos.* **2023**, *140*, 105104. [CrossRef]
27. Fernandes, B.; Khodeir, M.; Perlot, C.; Carré, H.; Mindeguia, J.-C.; La Borderie, C. Durability of concrete made with recycled concrete aggregates after exposure to elevated temperatures. *Mater. Struct.* **2023**, *56*, 25. [CrossRef]
28. Medina, D.F.; Martínez, M.C.H.; Medina, N.F.; Hernández-Olivares, F. Durability of rubberized concrete with recycled steel fibers from tyre recycling in aggresive enviroments. *Constr. Build. Mater.* **2023**, *400*, 132619. [CrossRef]
29. Lin, Z. *Cementing Materials Science*; Wuhan University of Technology Press: Wuhan, China, 2014; pp. 110–125. (In Chinese)

Disclaimer/Publisher's Note: The statements, opinions and data contained in all publications are solely those of the individual author(s) and contributor(s) and not of MDPI and/or the editor(s). MDPI and/or the editor(s) disclaim responsibility for any injury to people or property resulting from any ideas, methods, instructions or products referred to in the content.

Article

Effect of Curing Conditions on the Hydration of MgO in Cement Paste Mixed with MgO Expansive Agent

Xuefeng Zhao [1,*], Zhongyang Mao [1], Xiaojun Huang [1], Penghui Luo [1], Min Deng [1,2] and Mingshu Tang [1,2,*]

[1] College of Materials Science and Engineering, Nanjing Tech University, Nanjing 211800, China; 5967@njtech.edu.cn (X.H.); 202162103025@njtech.edu.cn (P.L.)
[2] State Key Laboratory of Materials-Oriented Chemical Engineering, Nanjing 211800, China
* Correspondence: 202061103120@njtech.edu.cn (X.Z.); tangmingshu@njtech.edu.cn (M.T.); Tel.: +86-136-0518-4865 (M.T.)

Abstract: Using the volume expansion generated by the hydration of the MgO expansive agent to compensate for the shrinkage deformation of concrete is considered to be an effective measure to prevent concrete shrinkage and cracking. Existing studies have mainly focused on the effect of the MgO expansive agent on the deformation of concrete under constant temperature conditions, but mass concrete in practical engineering experiences a temperature change process. Obviously, the experience obtained under constant temperature conditions makes it difficult to accurately guide the selection of the MgO expansive agent under actual engineering conditions. Based on the C50 concrete project, this paper mainly investigates the effect of curing conditions on the hydration of MgO in cement paste under actual variable temperature conditions by simulating the actual temperature change course of C50 concrete so as to provide a reference for the selection of the MgO expansive agent in engineering practice. The results show that temperature was the main factor affecting the hydration of MgO under variable temperature curing conditions, and the increase in the temperature could obviously promote the hydration of MgO in cement paste, while the change in the curing methods and cementitious system had an effect on the hydration of MgO, though this effect was not obvious.

Keywords: MgO expansive agent; variable temperature curing; hydration degree; curing condition

---

Citation: Zhao, X.; Mao, Z.; Huang, X.; Luo, P.; Deng, M.; Tang, M. Effect of Curing Conditions on the Hydration of MgO in Cement Paste Mixed with MgO Expansive Agent. *Materials* **2023**, *16*, 4032. https://doi.org/10.3390/ma16114032

Academic Editor: Weiting Xu

Received: 6 May 2023
Revised: 19 May 2023
Accepted: 23 May 2023
Published: 28 May 2023

Copyright: © 2023 by the authors. Licensee MDPI, Basel, Switzerland. This article is an open access article distributed under the terms and conditions of the Creative Commons Attribution (CC BY) license (https://creativecommons.org/licenses/by/4.0/).

1. Introduction

Although concrete has been used as a building material for a long time, some basic problems have not been completely solved. The shrinkage deformation of concrete leading to the decline of durability is one of the most important problems [1–4]. The shrinkage deformation of concrete is a non-external deformation caused by the combined action of physics and chemistry. In actual structures, this volume deformation is often limited by external constraints, resulting in stresses within the concrete [5]. The cracking of concrete not only reduces the mechanical properties of the structure and affects the beauty of the concrete building but also provides a convenient channel for harmful ions to enter the concrete interior, which could further affect the durability of the concrete and reduce the service life of the concrete structures [6]. When the durability of the concrete decreased seriously, a large amount of production materials was spent on either the maintenance or even the reconstruction of the concrete structure, which is contrary to the concept of energy conservation and the emissions reduction proposed today.

In order to reduce the negative effects caused by the shrinkage cracking of concrete, researchers have attempted to use a variety of anti-cracking measures in engineering practice and achieved better anti-cracking effects. The anti-cracking measures for concrete mainly were considered in the following three directions: (1) Reduce the shrinkage deformation generated by the concrete itself: For example, covering the surface of concrete in a timely manner after it has been formed and spraying water on the surface periodically as a means of maintaining surface moisture to reduce the risk of surface cracking [7]; reducing the peak temperature of concrete which decreases the impact of the temperature drop shrinkage (for mass concrete, the value of the temperature rise of concrete can be more than 40 °C, plus the initial temperature value of concrete can be up to 75 °C~85 °C, so the concrete produces a huge shrinkage stress in the cooling stage). The means of reducing the peak temperature of concrete include embedding cooling water pipes inside the mass concrete, taking away part of the heat by cooling circulating water [8,9], using low and medium-heat Portland cement to reduce the heat release of cement hydration, replacing part of the cement with mineral admixtures such as fly ash to reduce the heat of hydration of cementitious [10,11], etc. The use of shrinkage reducers to reduce the tension of the internal capillaries of concrete can provide a means of reducing shrinkage due to a lower relative humidity inside the concrete [12–14]. The use of internal curing agents (water-saturated low-density aggregates and highly absorbent polymeric materials, etc.) to slow down the drying shrinkage during the hydration of cement by releasing water stored inside the curing agent [15–17], etc. (2) Improving the cracking resistance of concrete by adding fibers to concrete increases its tensile strength by virtue of the tensile capacity of the incorporated fibers and inhibits the possibility of internal microscopic cracks that can develop into macroscopic cracks [18–21]. (3) Adding an expansion component to concrete by adding expansion sources to concrete can compensate for the shrinkage deformation of concrete through the volume expansion generated by the expansion component during hydration, thereby alleviating or avoiding the generation of shrinkage cracks [22–24].

The above anti-cracking means have proved their effectiveness through long-term practice, but the shortcomings of some anti-cracking measures have also been found in the process of this practice. For the mitigation of the thermal shrinkage of mass concrete, the use of the cooling water pipe can play a role in reducing thermal shrinkage, but there are problems, such as complex construction, the high cost of cooling facilities and a long construction cycle [8,25]. In addition, adding fiber to concrete can enhance the tensile properties of the concrete, but there are also some problems such as the addition of fiber which leads to the poor fluidity and workability of concrete. Additionally, how to make the fiber uniform distribution in the concrete is also a problem to be solved [26]. Relatively speaking, adding an expansive agent to concrete is an economical, simple and effective control method [23].

At present, the common expansive agent mainly includes a sulfur aluminate type expansive agent, calcium oxide expansive agent and magnesium oxide expansive agent. The sulfur aluminate expansive agent hydration requires a large amount of water, and the thermal stability of sulfur aluminate hydration products is poor, meaning it can undergo dehydration decomposition at temperatures higher than 70 °C [24,27]. The hydration rate of the expansion of the calcium oxide expansive agent is very fast, and it is not easy to control [8,28]. Compared to calcium oxide and sulfur aluminate type expansive agents, the hydration expansion of the MgO expansive agent requires less water, and the physical and chemical properties of hydration products are more stable. Moreover, the expansion rate of magnesium oxide can also be regulated and designed, which makes the application of the MgO expansive agent in engineering practice more and more common [29,30].

Although the MgO expansive agent has a good effect on preventing concrete cracking, existing studies mainly focus on the influence of the MgO expansive agent on concrete deformation under constant temperature conditions, while engineering mass concrete experiences a variable temperature process, especially for high-strength concrete, due to a large amount of cementitious material, with higher temperature rises. In order to provide some guidance for the selection of the MgO expansive agent under actual engineering conditions based on C50 concrete engineering, this paper mainly studies the effect of curing conditions on MgO hydration in a cement paste-mixed MgO expansive agent under the actual variable temperature condition. In addition, the hydration of MgO in cement paste formed with two kinds of cementitious systems that are commonly used in engineering was compared.

2. Materials and Methods

2.1. Materials

The raw materials used in this experiment included Portland cement, secondary fly ash, S95 slag powder and the MgO expansive agent (i.e., MEA). Among them, Portland cement was P·II52.5 cement produced by Jiangnan Onoda Cement Co., Ltd., Nanjing, China, and secondary fly ash and S95 mineral powder was provided by Nanjing Pudi Concrete Company, Nanjing, China. Four kinds of MgO expansive agents were provided by Wuhan Sanyuan Special Building Materials Company and Jiangsu Sobute Company. According to the citric acid method [31], the reactivity values of four kinds of MgO expansive agent were 120s, 180s, 240s and 330s. Additionally, they were named MEA-120, MEA-180, MEA-240 and MEA-330, respectively. Table 1 shows the chemical composition of Portland cement, secondary fly ash, S95 slag powder and the four kinds of magnesium oxide expansive agent used in the experiment.

Table 1. Chemical compositions of raw materials.

Item	Chemical Components (%)								
	CaO	SiO$_2$	Al$_2$O$_3$	Fe$_2$O$_3$	MgO	SO$_3$	Na$_2$O	K$_2$O	LOSS
Portland cement	65.32	18.55	3.95	3.41	1.01	2.78	0.72	0.17	2.88
Fly ash	4.07	50.53	31.65	4.48	0.91	1.32	0.67	1.26	2.77
Slag	38.00	33.72	17.74	0.77	6.35	1.04	0.41	0.40	−0.72
MEA-120	1.89	3.38	0.54	0.66	85.23	0.92	-	-	3.12
MEA-180	3.90	5.17	0.64	0.68	85.57	0.75	-	-	2.97
MEA-240	1.88	4.07	0.85	0.78	90.45	0.03	-	-	1.53
MEA-330	4.37	5.82	0.71	0.61	85.71	0.60	-	-	1.88

2.2. Methods

2.2.1. Simulation of the Temperature Change Process of Mass Concrete

In this experiment, the simulation of the temperature change process of mass concrete was based on the actually measured temperature data of C50 mass concrete in engineering, and the curing box and external temperature acquisition module were used to realize the stage temperature change.

The temperature simulation process was as follows: the temperature change process of concrete was broken down into several small temperature change stages. Additionally, the temperature was adjusted every 4 h during the heating phase until it increased to the maximum temperature; the temperature was adjusted every 8 h during the cooling phase until it dropped to an ambient temperature. The heating and cooling rate of the temperature was determined according to the concrete temperature data measured in the engineering. Figure 1 shows the temperature variation process inside two groups of concrete walls that were measured in a C50 mass concrete engineering project. As can be seen from Figure 1, the internal temperature of concrete reached its maximum value around 24–48 h (taking its time when concrete began to be poured as the initial zero point), and the internal temperature of concrete was basically consistent with the ambient temperature at 14 d, i.e., the cooling process was basically over. The simulation of the temperature variation process in this study was based on the temperature variation in the data of mass concrete obtained from engineering practice (as shown in Figure 1), and Figure 2 shows the variable temperature curing environment with peak temperatures at 65 °C and 85 °C, which were simulated by a curing box based on the temperature change data obtained in engineering practice.

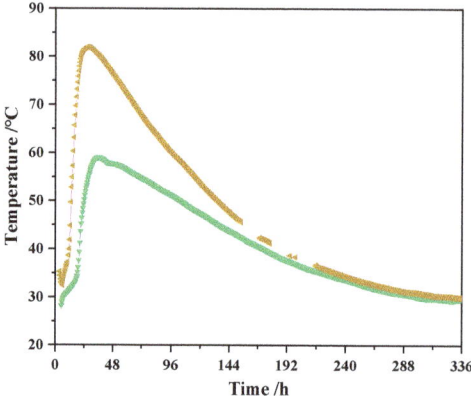

Figure 1. Temperature variation process of mass concrete measured.

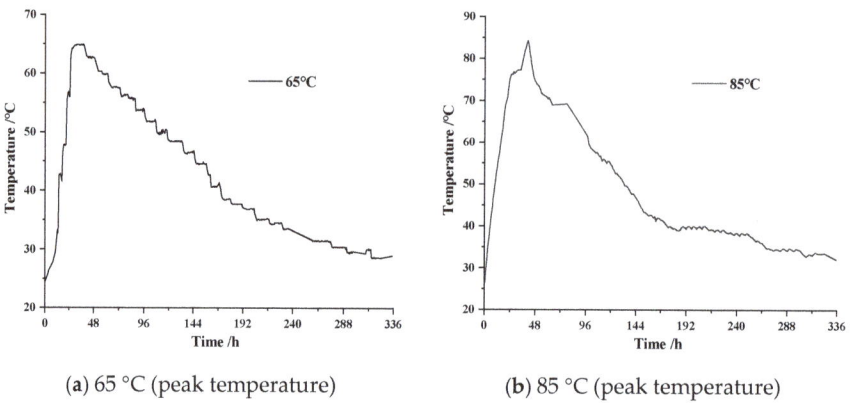

(**a**) 65 °C (peak temperature)   (**b**) 85 °C (peak temperature)

Figure 2. Variable temperature curing environment simulated by curing box: (**a**) 65 °C (peak temperature); (**b**) 85 °C (peak temperature).

2.2.2. Preparation and Curing of Cement Pastes

The tests involved a total of two cementitious systems, and the raw material formulation of the formed cement paste is shown in Table 2. The MEA content in the two cementitious systems accounted for 8% of the total mass of the cementitious materials. The cement paste test block used to test the hydration of MgO was molded in Φ25 mm × 30 mm columnar rigid PVC molds with a water-cement ratio of 0.32. Before forming, the raw materials were evenly mixed by a mixer, and the evenly mixed cement paste was put into the test mold by a cement paste purifying mixer with water for mixing. Then, the test mold was placed on a shaking table for 60 s to eliminate the air inside the paste. The prepared cement paste specimens were maintained in the curing environment shown in Figure 2 to observe the hydration of MgO in the cement paste.

Table 2. Mix proportion of cement paste/g.

System	Cement	Fly Ash	Slag	MEA	Water
a	66	18	8	8	32
b	74	18	-	8	32

The experiment involved two curing methods: water curing and non-wet curing. The prepared cement paste specimens were cured directly with a mold (the mold could be sealed with a cover), and the mold was wrapped with plastic wrap to isolate water so as to realize the not-wet curing. The specimens of the cement paste cured in water were removed from the mold and placed in water for curing after 20 h of curing with the mold. In addition, paste specimens with curing ages of 1 d, 3 d, 7 d and 14 d were selected to complete the characterization of MgO's hydration degree under the whole temperature course (take the preparation time of ready cement paste as the initial zero point).

2.2.3. Determination of the Remaining MgO Content in Cement Paste

The K-value method (XRD) was used to determine the content of MgO in the cement paste for hydration. ZnO was chosen as the internal standard substance (the dosage is 2 wt.%). The scanning range was 35°–45°, and the scanning speed was 1°/min. ZnO and MgO powders were, respectively, weighed according to the mass ratio of 1:1, and the powders were ground with an agate mortar to obtain a homogeneous mixed powder sample. The milled powder sample was subjected to X-ray diffraction analysis, and the integrated area of the strongest diffraction peaks of ZnO and MgO in the diffraction pattern was calculated using MDI Jade software. The ratio of the integrated area of the strongest diffraction peaks of ZnO and MgO was the K value, which was calculated to be 0.57. The position of the strongest diffraction peak of ZnO was d_{101} = 2.47, and the position of the strongest diffraction peak of MgO was d_{200} = 2.11. The remaining content of MgO in the cement paste was calculated by Equation (1):

$$\omega_{MgO} = K \frac{(1-\omega_{ZnO})}{\omega_{ZnO}} \frac{I_{ZnO}}{I_{MgO}} \tag{1}$$

where ω_{MgO} is the remaining content of MgO in the cement paste. ω_{ZnO} is the doping amount of the internal standard substance ZnO. I_{ZnO} is the integrated intensity of the strongest diffraction peak of the internal standard substance ZnO, and I_{MgO} is the integrated intensity of the strongest diffraction peak of MgO. K is the characteristic constant (K = 0.57).

3. Results and Discussion

3.1. Effect of Curing Temperature on MgO Hydration in Cement Paste

In this section, the effect of the curing temperature on MgO hydration in cement paste was studied. The cement paste was prepared according to the raw material ratio of "system a" in Table 2. Figure 3 shows the hydration process of MgO when mixed with four types of MEA cement paste under a 65 °C variable temperature with water curing conditions. It can be seen from Figure 3 that the changes in the MgO content in the cement paste mixed with four types of MEA generally showed a trend of first fast and then slow. The hydration rate of MgO in the cement paste was the fastest before 1 d, and then the hydration rate of MgO in the cement paste began to gradually slow down. When the curing age was 1 d, the content of MgO in the cement paste mixed with MEA-120, MEA-180, MEA-240 and MEA-330 was reduced by 3.48 wt.%, 2.85 wt.%, 2.45 wt.% and 2.33 wt.%, respectively. In contrast, the content of MgO in the cement paste mixed with MEA-120, MEA-180, MEA-240 and MEA-330 was reduced by 0.99 wt.%, 0.81 wt.%, 0.88 wt.% and 0.63 wt.% at the curing age from 1 d to 3 d, respectively. In addition, the higher the activity of MEA, the faster the hydration at the early stage, and the lower MgO remained in the cement paste during the cooling stage. The cement pastes with MEA-120 had the lowest residual MgO content throughout the process, while the cement paste with MEA-330 had the highest residual MgO content throughout the process. The residual MgO content of the cement paste with the other two activities always lay between them.

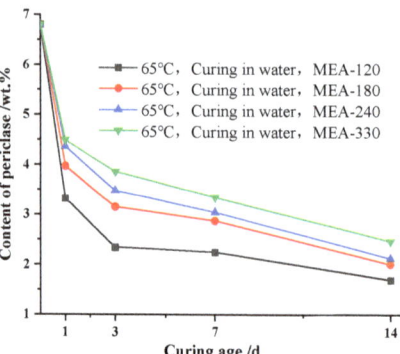

Figure 3. Content of periclase for different ages in cement pastes with MEA cured in water at 65 °C variable temperature condition.

Figure 4 shows the hydration process of MgO mixed with four MEA cement pastes under 85 °C variable temperature water curing conditions. As can be seen from Figure 4, the trend in the change in the MgO content in cement paste mixed with four kinds of MEA at 65 °C and 85 °C and under a variable temperature water curing condition was basically similar. The hydration rate of MgO in the cement paste was the fastest at 1 d, and then the hydration rate started to decrease gradually. However, compared to the 65 °C variable temperature water curing condition, the residual content of MgO in the cement paste under an 85 °C variable temperature water curing condition at a curing age of 1 d was significantly reduced. Moreover, the reaction rate of MgO in the cement paste became slower during the cooling stage, and the content of MgO in the cement paste of multiple MEA-doped groups appeared as a "plateau period"., i.e., the content of MgO remained essentially unchanged.

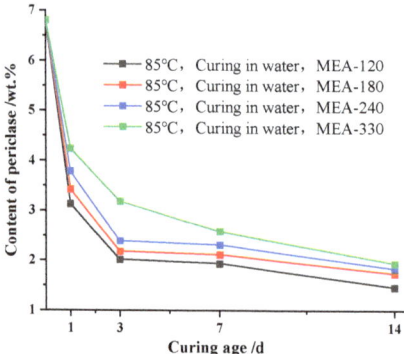

Figure 4. Content of periclase for different ages in cement pastes with MEA cured in water at 85 °C with variable temperature conditions.

Figure 5 shows the comparison of the remaining MgO content of the cement paste mixed with the same active MEA at 65 °C and 85 °C under variable temperature water curing conditions. By comparison, it was found that the residual content of MgO in the cement paste mixed with MEA-120, MEA-180, MEA-240 and MEA-330 at the curing age of 1 d under an 85 °C variable temperature and the water curing condition is reduced by 0.2 wt.%, 0.54 wt.%, 0.57 wt.% and 0.24 wt.%, respectively, compared with that under a 65 °C variable temperature water curing condition. Moreover, the MgO content in the cement paste mixed with MEA-120 under a 65 °C variable temperature water curing condition and MEA-120, MEA-180 and MEA-240 under an 85 °C variable temperature water curing condition showed a "plateau period" during the cooling stage (3–14 d). This was not significantly observed for other cement pastes when mixed with other active MEAs under the same curing condition. In comparison, it was found that the MgO content in the cement paste of these four groups started to show signs of a "plateau" mostly at the age of 3 d (i.e., the cooling stage). At this time, the MgO content in the cement paste was mixed with MEA-120 under a 65 °C variable temperature water curing condition, MEA-120, MEA-180 and MEA-240 was under an 85 °C variable temperature water curing condition and were 2.33 wt.%, 2.01 wt.%, 2.17 wt.% and 2.38 wt.%, respectively. The other four groups of cement pastes without this situation had a higher MgO content at a curing age of 3 d compared to the MgO content values at the time of plateauing. In addition, it could also be found that the increase in the curing temperature had a slightly different effect on the promotion of the hydration reaction of different active MEAs. The low activity of MEA seemed to be more sensitive to the temperature, and the increase in the temperature was more effective when promoting the hydration of low-activity MEA. This may also be due to the rapid reaction of the highly reactive MEA, resulting in a premature approach to the "plateau content" of MgO in the cement paste mixed with a highly reactive MEA.

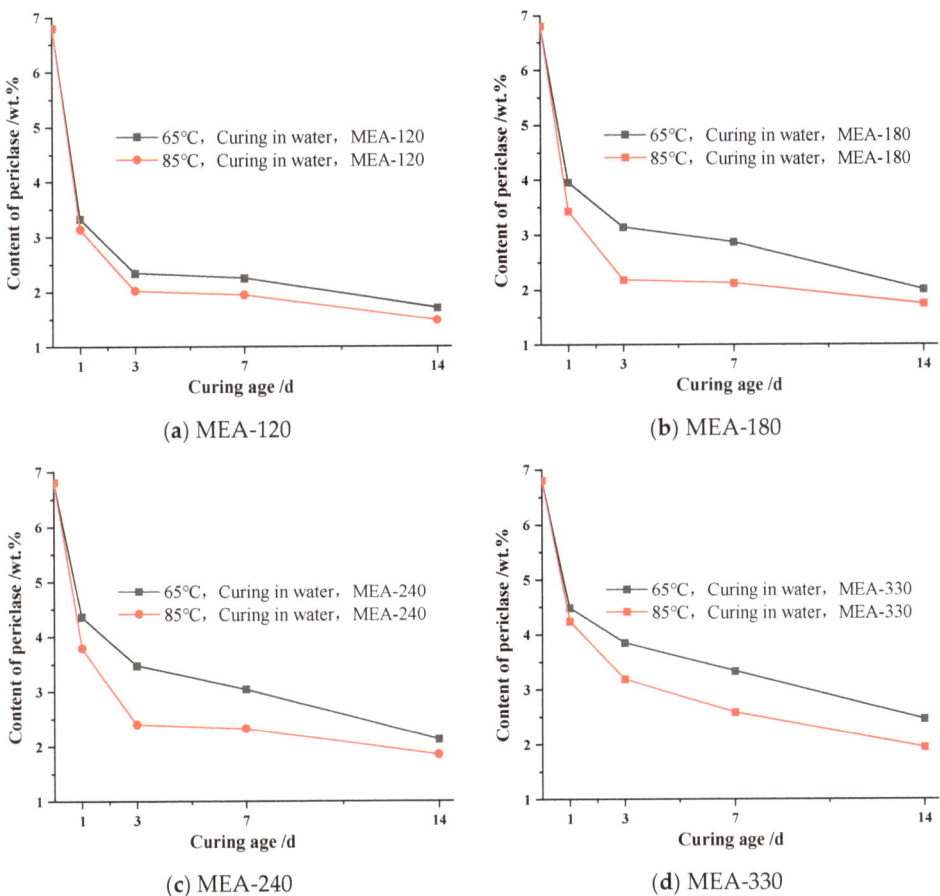

Figure 5. Content of periclase for different ages in cement pastes mixed with the same MEA cured in water at 65 °C, 85 °C variable temperature: (**a**) MEA-120; (**b**) MEA-180; (**c**) MEA-240; (**d**) MEA-330.

3.2. Effect of Curing Methods on the Hydration of MgO in Cement Paste

In order to clarify the effect of water curing and non-wet curing on MgO hydration in the cement paste under variable temperature conditions, the hydration of MgO in the cement paste was compared under two curing methods according to "system a" forming cement paste. Figure 6 shows the comparison of MgO residual content in cement paste mixed with the same active MEA by different curing methods under variable temperature conditions of 65 °C and 85 °C.

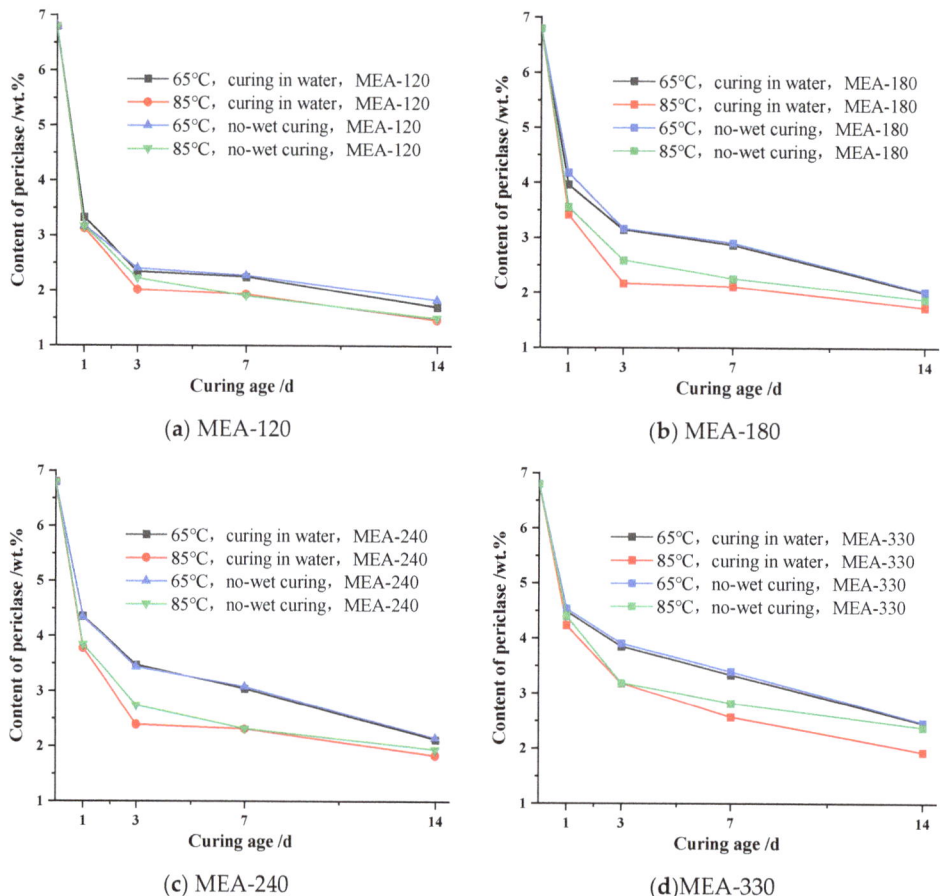

Figure 6. Content of periclase for different ages in cement pastes mixed with the same MEA under different curing conditions: (**a**) MEA-120; (**b**) MEA-180; (**c**) MEA-240; (**d**) MEA-330.

As can be seen in Figure 6, for cement paste mixed with the same active MEA curing in the same variable temperature environment, the effect of non-wet curing and water curing on the hydration of MgO in the cement paste was not as great as the effect of temperature on the hydration of MgO. The effect of moisture on MgO hydration in the MEA mixed cement paste was mainly reflected in the period before the curing age of 3 d, and the promotion effect of moisture on MgO hydration in the cement paste in the subsequent process seemed not to be not obvious. It can be seen from the changing curve of MgO content in the cement paste underwater with non-wet curing at a 65 °C variable temperature that the difference between non-wet and water curing methods on MgO hydration was very small at the curing age of 1–3 d, and the remaining MgO content in the cement paste under the two curing methods of MEA-120, MEA-180, MEA-240 and MEA-330 was basically the same. In addition, in the subsequent cooling process, no significant difference was observed in the content of MgO on the paste under two curing conditions, even with the different curing methods. As can be seen from the variation curves of the content of MgO in cement paste under 85 °C variable temperature conditions for water and non-wet curing, compared with non-wet curing, the remaining content of MgO in the cement paste mixed with MEA under water curing significantly decreased at this stage before the curing age of 3 d, where no significant difference was found in the change in the MgO content in the paste under the two curing methods of water and non-wet curing in the cooling stage after 3 d.

When comparing the hydration process of MgO in the cement paste in Figure 6c, it was found that the content of MgO in the cement paste mixed with MEA was significantly reduced when the temperature peak of variable temperature curing increased from 65 °C to 85 °C under the same non-wet condition. Compared to 65 °C variable temperature non-wet curing, the residual content of MgO in the cement paste at 85 °C variable temperature non-wet curing ages of 1 d, 3 d, 7 d and 14 d decreased by 0.50 wt.%, 0.69 wt.%, 0.76 wt.% and 0.21 wt.%, respectively. However, under the same 65 °C variable temperature curing condition, when the curing method changed from non-wet curing to water curing, there was no significant difference in the content of MgO in the cement paste when mixed with MEA-240 during the whole age. In addition, it could also be found that the MgO content in the cement paste mixed with MEA-240 was basically the same in this stage before the maintenance age of 3 d under the condition of water curing and non-wet curing at a 65 °C variable temperature curing. However, when the temperature peak of variable temperature curing from 65 °C increased to 85 °C, compared with the non-wet curing, the content of MgO in the water-curing cement paste mixed with MEA-240 decreased significantly by 0.35 wt.% at the curing age of 3 d. The variation in the MgO content in the cement paste when mixed with other active MEA in Figure 6 was also similar to the results shown in Figure 6c.

3.3. Effect of Cementitious System on MgO in Cement Paste

In order to clarify the effect of the cementitious system on MgO hydration in the cement paste under variable temperatures and non-wet curing conditions, the hydration of MgO in the cement paste was formed according to the cementitious system a and b was compared. Figure 7 shows the comparison of the MgO residual content in cement paste mixed with the same active MEA under the condition of variable temperature and non-wet curing at 65 °C and 85 °C.

As can be seen from Figure 7, for the cement paste formed according to the cementing systems a and b and mixed with the same active MEA, there was little difference in the variation in the MgO content in the cement paste under the same curing condition. The content of MgO in cement paste formed according to cementing system b was lower than that of the cement paste formed according to system a at all ages. This may be caused by the addition of an 8 wt.% S95 slag powder in system a to replace part of the cement without changing the yield of fly ash. According to Figure 7d, under variable temperature and non-wet curing at 65 °C, the residual content of MgO in the cement paste was formed according to the cementitious system a and was 0.12 wt.%, 0.14 wt.%, 0.10 wt.% and 0.04 wt.% higher than that in the cementitious system b at the age of 1 d, 3 d, 7 d and 14 d, respectively. Therefore, it can be considered that the addition of the 8 wt.% slag powder with the same fly ash content could inhibit the hydration of MgO in the cement paste; however, the inhibition effect was not obvious.

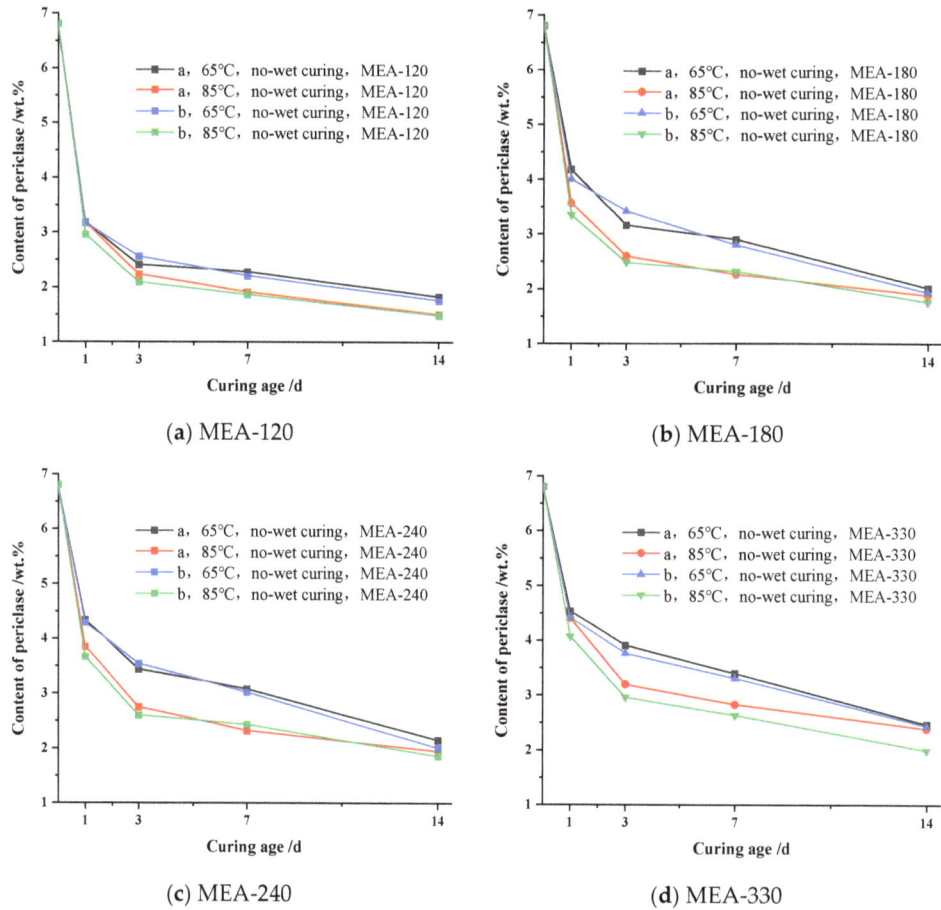

Figure 7. Content of periclase at different ages in cement pastes prepared by Cementitious system a and system b cured in non-wet conditions at different variable temperatures: (**a**) MEA-120; (**b**) MEA-180; (**c**) MEA-240; (**d**) MEA-330.

In addition, by comparing MgO hydration in the cement paste after the cementitious system and temperature change in Figure 7, it was found that the effect of temperature variation on MgO hydration in the cement paste was obviously greater than that of the cementitious system., i.e., the temperature was still the main factor affecting MgO hydration in cement paste.

4. Conclusions

1. Under variable temperature curing conditions with a peak temperature of 65 °C and 85 °C, the hydration of MgO in cement paste presented a trend that produced first fast and then slowly changed. The MgO of the cement paste mixed with highly active MEA was nearly half hydrated at the curing age of 1 d. After that, the change in the remaining content of MgO in the cement slurry slowed down significantly.
2. When the peak temperature of variable temperature curing increased from 65 °C to 85 °C, it accelerated the hydration rate of MgO in the cement paste in the early stage (1–3 d), and the increase in the curing temperature led to a plateau period for the change in the MgO content in cement paste. It was observed that when there

was a plateau of the MgO content change, MgO in the cement paste was nearly 70% hydrated (remaining content above 2.0 wt.%).
3. When the curing temperature and the cementitious system were the same, the effect of curing methods on the hydration of MgO in the cement paste was mainly reflected in the early stage (1–3 d). With the increase in the curing age, the difference in the MgO content caused by curing methods gradually decreased. In addition, the increase in the variable curing temperature could aggravate the effect of curing methods on MgO hydration in the cement paste.
4. By comparing the effects of temperature, the curing methods and the cementitious system on MgO hydration in the cement paste, it was found that temperature was the main factor affecting the hydration of MgO. The increase in the temperature could significantly promote the hydration of MgO in cement paste, under the condition of non-wet curing, compared to 65 °C variable temperature non-wet curing. The residual content of MgO in cement paste at a 85 °C variable temperature with non-wet curing ages 1 d, 3 d, 7 d and 14 d decreased by 0.50 wt.%, 0.69 wt.%, 0.76 wt.% and 0.21 wt.%, respectively; the other two had effects on the hydration of MgO, but this effect was not obvious.

Author Contributions: Conceptualization, M.D. and M.T.; data curation, X.Z. and X.H.; writing—original draft preparation, X.Z.; writing—review and editing, M.D., P.L. and Z.M. All authors have read and agreed to the published version of the manuscript.

Funding: This work was supported by the Open Fund project of the National Laboratory for High Performance Civil Engineering Materials (2021CEM012) and the Priority Academic Program Development of Jiangsu Higher Education Institutions (PAPD).

Institutional Review Board Statement: Not applicable.

Informed Consent Statement: Not applicable.

Data Availability Statement: The data presented in this study are available on request from the corresponding author.

Acknowledgments: The authors gratefully acknowledge the assistance from Zhongyang Mao and Xiaojun Huang from NJTECH, and the staff from State Key Laboratory of Materials-Oriented Chemical Engineering.

Conflicts of Interest: The authors declare no conflict of interest.

References

1. Kayondo, M.; Combrinck, R.; Boshoff, W.P. State-of-the-Art Review on Plastic Cracking of Concrete. *Constr. Build. Mater.* **2019**, *225*, 886–899. [CrossRef]
2. Tran, N.P.; Gunasekara, C.; Law, D.W.; Houshyar, S.; Setunge, S.; Cwirzen, A. A Critical Review on Drying Shrinkage Mitigation Strategies in Cement-Based Materials. *J. Build. Eng.* **2021**, *38*, 102210. [CrossRef]
3. Wu, L.M.; Farzadnia, N.; Shi, C.J.; Zhang, Z.H.; Wang, H. Autogenous Shrinkage of High Performance Concrete: A Review. *Constr. Build. Mater.* **2017**, *149*, 62–75. [CrossRef]
4. Yodsudjai, W.; Wang, K.J. Chemical Shrinkage Behavior of Pastes Made with Different Types of Cements. *Constr. Build. Mater.* **2013**, *40*, 854–862. [CrossRef]
5. Neville, A.M. *Properties of Concrete*; Longman: London, UK, 1995; Volume 4.
6. Shi, X.M.; Xie, N.; Fortune, K.; Gong, J. Durability of Steel Reinforced Concrete in Chloride Environments: An Overview. *Constr. Build. Mater.* **2012**, *30*, 125–138. [CrossRef]
7. Huo, X.S.; Wong, L.U. Experimental Study of Early-Age Behavior of High Performance Concrete Deck Slabs Under Different Curing Methods. *Constr. Build. Mater.* **2006**, *20*, 1049–1056. [CrossRef]
8. Mo, L.; Deng, M.; Tang, M.; Al-Tabbaa, A. MgO Expansive Cement and Concrete in China: Past, Present and Future. *Cem. Concr. Res.* **2014**, *57*, 1–12. [CrossRef]
9. Zhong, R.; Hou, G.P.; Qiang, S. An Improved Composite Element Method for the Simulation of Temperature Field in Massive Concrete with Embedded Cooling Pipe. *Appl. Therm. Eng.* **2017**, *124*, 1409–1417. [CrossRef]
10. Bouasker, M.; Mounanga, P.; Turcry, P.; Loukili, A.; Khelidj, A. Chemical Shrinkage of Cement Pastes and Mortars at Very Early Age: Effect of Limestone Filler and Granular Inclusions. *Cem. Concr. Comp.* **2008**, *30*, 13–22. [CrossRef]

11. Tertnkhajornkit, P.; Nawa, T.; Nakai, M.; Saito, T. Effect of Fly Ash on Autogenous Shrinkage. *Cem. Concr. Res.* **2005**, *35*, 473–482. [CrossRef]
12. Bentz, D.P.; Geiker, M.R.; Hansen, K.K. Shrinkage-Reducing Admixtures and Early-Age Desiccation in Cement Pastes and Mortars. *Cem. Concr. Res.* **2001**, *31*, 1075–1085. [CrossRef]
13. Collepardi, M.; Borsoi, A.; Collepardi, S.; Olagot, J.; Troli, R. Effects of Shrinkage Reducing Admixture in Shrinkage Compensating Concrete Under Non-Wet Curing Conditions. *Cem. Concr. Comp.* **2005**, *27*, 704–708. [CrossRef]
14. Yang, G.; Wu, Y.K.; Li, H.; Gao, N.X.; Jin, M.; Hu, Z.L.; Liu, J.P. Effect of Shrinkage-Reducing Polycarboxylate Admixture on Cracking Behavior of Ultra-High Strength Mortar. *Cem. Concr. Comp.* **2021**, *122*, 104117. [CrossRef]
15. Liu, J.H.; Shi, C.J.; Ma, X.W.; Khayat, K.H.; Zhang, J.; Wang, D.H. An Overview on the Effect of Internal Curing on Shrinkage of High Performance Cement-Based Materials. *Constr. Build. Mater.* **2017**, *146*, 702–712. [CrossRef]
16. Masum, A.; Manzur, T. Delaying Time to Corrosion Initiation in Concrete Using Brick Aggregate as Internal Curing Medium Under Adverse Curing Conditions. *Constr. Build. Mater.* **2019**, *228*, 116772. [CrossRef]
17. Xu, F.M.; Lin, X.S.; Zhou, A.N. Performance of Internal Curing Materials in High-Performance Concrete: A Review. *Constr. Build. Mater.* **2021**, *311*, 125250. [CrossRef]
18. Barluenga, G.; Hernandez-Olivares, F. Cracking Control of Concretes Modified with Short AR-Glass Fibers at Early Age. Experimental Results on Standard Concrete and Scc. *Cem. Concr. Res.* **2007**, *37*, 1624–1638. [CrossRef]
19. Bertelsen, I.; Ottosen, L.M.; Fischer, G. Influence of Fiber Characteristics on Plastic Shrinkage Cracking in Cement-Based Materials: A Review. *Constr. Build. Mater.* **2020**, *230*, 116769. [CrossRef]
20. Fang, C.F.; Ali, M.; Xie, T.Y.; Visintin, P.; Sheikh, A.H. The Influence of Steel Fiber Properties on the Shrinkage of Ultra-High Performance Fibre Reinforced Concrete. *Constr. Build. Mater.* **2020**, *242*, 117993. [CrossRef]
21. Shen, D.J.; Wen, C.Y.; Zhu, P.F.; Wu, Y.H.; Yuan, J.J. Influence of Barchip Fiber on Early-Age Autogenous Shrinkage of High Strength Concrete. *Constr. Build. Mater.* **2020**, *256*, 119223. [CrossRef]
22. Konik, Z.; Malolepszy, J.; Roszczynialski, W.; Stok, A. Production of Expansive Additive to Portland Cement. *J. Eur. Ceram. Soc.* **2007**, *27*, 605–609. [CrossRef]
23. Mo, L.; Deng, M.; Wang, A. Effects of Mgo-Based Expansive Additive on Compensating the Shrinkage of Cement Paste Under Non-Wet Curing Conditions. *Cem. Concr. Compos.* **2012**, *34*, 377–383. [CrossRef]
24. Nagataki, S.; Gomi, H. Expansive Admixtures (Mainly Ettringite). *Cem. Concr. Comp.* **1998**, *20*, 163–170. [CrossRef]
25. Lu, X.C.; Chen, B.F.; Tian, B.; Li, Y.B.; Lv, C.C.; Xiong, B.B. A New Method for Hydraulic Mass Concrete Temperature Control: Design and Experiment. *Constr. Build. Mater.* **2021**, *302*, 124167. [CrossRef]
26. Zhao, C.G.; Wang, Z.Y.; Zhu, Z.Y.; Guo, Q.Y.; Wu, X.R.; Zhao, R.D. Research on Different Types of Fiber Reinforced Concrete in Recent Years: An Overview. *Constr. Build. Mater.* **2023**, *365*, 130075. [CrossRef]
27. Collepardi, M. A State-of-the-Art Review on Delayed Ettringite Attack on Concrete. *Cem. Concr. Comp.* **2003**, *25*, 401–407. [CrossRef]
28. Zhao, H.T.; Jiang, K.D.; Di, Y.F.; Xu, W.; Li, W.; Tian, Q.; Liu, J.P. Effects of Curing Temperature and Superabsorbent Polymers on Hydration of Early-Age Cement Paste Containing a CaO-Based Expansive Additive. *Mater. Struct.* **2019**, *52*, 108. [CrossRef]
29. Mo, L.; Deng, M.; Tang, M. Effects of Calcination Condition on Expansion Property of MgO-Type Expansive Agent Used in Cement-Based Materials. *Cem. Concr. Res.* **2010**, *40*, 437–446. [CrossRef]
30. Salomao, R.; Bittencourt, L.; Pandolfelli, V.C. A Novel Approach for Magnesia Hydration Assessment in Refractory Castables. *Ceram. Int.* **2007**, *33*, 803–810. [CrossRef]
31. Mo, L.W.; Fang, J.W.; Hou, W.H.; Ji, X.K.; Yang, J.B.; Fan, T.T.; Wang, H.L. Synergetic Effects of Curing Temperature and Hydration Reactivity of MgO Expansive Agents on their Hydration and Expansion Behaviours in Cement Pastes. *Constr. Build. Mater.* **2019**, *207*, 206–217. [CrossRef]

Disclaimer/Publisher's Note: The statements, opinions and data contained in all publications are solely those of the individual author(s) and contributor(s) and not of MDPI and/or the editor(s). MDPI and/or the editor(s) disclaim responsibility for any injury to people or property resulting from any ideas, methods, instructions or products referred to in the content.

Article

Study on the Effect of PVAc and Styrene on the Properties and Microstructure of MMA-Based Repair Material for Concrete

Zemeng Guo, Lingling Xu *, Shijian Lu, Luchao Yan, Zhipeng Zhu and Yang Wang

College of Materials Science and Engineering, Nanjing Tech University, Nanjing 211816, China; 202061103028@njtech.edu.cn (Z.G.); 202161103065@njtech.edu.cn (S.L.); 202161203204@njtech.edu.cn (L.Y.); 202161103006@njtech.edu.cn (Z.Z.); 202261103012@njtech.edu.cn (Y.W.)
* Correspondence: xll@njtech.edu.cn; Tel.: +86-139-5184-6173

Abstract: Methyl methacrylate (MMA) material is considered to be a suitable material for repairing concrete crack, provided that its large volume shrinkage during polymerization is resolved. This study was dedicated to investigating the effect of low shrinkage additives polyvinyl acetate and styrene (PVAc + styrene) on properties of the repair material and further proposes the shrinkage reduction mechanism based on the data of FTIR spectra, DSC testing and SEM micrographs. The results showed that PVAc + styrene delayed the gel point during the polymerization, and the formation of two-phase structure and micropores compensated for the volume shrinkage of the material. When the proportion of PVAc + styrene was 12%, the volume shrinkage could be as low as 4.78%, and the shrinkage stress was reduced by 87.4%. PVAc + styrene improved the bending strength and fracture toughness of most ratios investigated in this study. When 12% PVAc + styrene was added, the 28 d flexural strength and fracture toughness of MMA-based repair material were 28.04 MPa and 92.18%, respectively. After long-term curing, the repair material added with 12% PVAc + styrene showed a good adhesion to the substrate, with a bonding strength greater than 4.1 MPa and the fracture surface appearing at the substrate after the bonding experiment. This work contributes to the obtaining of a MMA-based repair material with low shrinkage, while its viscosity and other properties also can meet the requirements for repairing microcracks.

Keywords: methyl methacrylate; concrete repair material; low shrinkage; shrinkage stress; mechanical properties

Citation: Guo, Z.; Xu, L.; Lu, S.; Yan, L.; Zhu, Z.; Wang, Y. Study on the Effect of PVAc and Styrene on the Properties and Microstructure of MMA-Based Repair Material for Concrete. *Materials* 2023, *16*, 3984. https://doi.org/10.3390/ma16113984

Academic Editor: Carola Esposito Corcione

Received: 16 April 2023
Revised: 11 May 2023
Accepted: 24 May 2023
Published: 26 May 2023

Copyright: © 2023 by the authors. Licensee MDPI, Basel, Switzerland. This article is an open access article distributed under the terms and conditions of the Creative Commons Attribution (CC BY) license (https:// creativecommons.org/licenses/by/ 4.0/).

1. Introduction

Concrete is one of the most commonly used building materials in the world, but cracks are always present in concrete owing to its shrinkage, external freeze-thaw, chemical erosion and other processes [1,2]. They can destroy the strength of the concrete and shorten its service life, and can even cause further structural damage if left untreated [3]. Therefore, it is necessary to repair cracks of concrete in time, whether from the perspective of economy or environmental protection [4].

At present, repair materials commonly used in the world can be divided into three types [5]: inorganic repair material, which is the most widely used material to repair cracks in the world, mainly including sulphoaluminate cement, magnesium phosphate cement and expansive cement [6–8]; organic repair material, which usually refers to polymer resins such as polyurethane, epoxy and acrylate [9–11]; and organic–inorganic composite repair material, which combines the advantages of organic material and inorganic material in the form of polymer-modified mortar/concrete, polymer-impregnated mortar/concrete and polymer mortar/concrete [12–14].

Grouting is a method that can be used to fill the cracks with appropriate seriflux to achieve the purpose of reinforcement and antiseepage, and it is widely used in the repair works of microcracks [15,16]. Penetration capability is an important evaluation criterion for the success of grouting; especially for those microcracks with a width less than 0.5 mm [17], the grouting performance of cement-based repair materials and organic–inorganic composite repair materials is very limited due to their particle size [18,19], while the organic repair materials with excellent fluidity are the best repair materials for microcracks [20]. Epoxy resin is a typical organic repair material, with high strength, excellent chemical resistance and good adhesive property. However, its relatively high viscosity makes it limited in repairing microcracks [21]. Furfural/acetone dilution system is usually used to reduce the viscosity of epoxy resin, but furfural is toxic and volatile, which will cause harm to human body and environment [22]. Its high brittleness and poor toughness also greatly limit its application [23]. Polyurethane is also a commonly used chemical grouting material, which has good impermeability and high strength [24]. However, the curing process of polyurethane requires reaction with water, so polyurethane is mostly used in the repair of water conservancy engineering [25]. In addition, polyurethane also needs diluent to reduce its viscosity [26]. Therefore, it is very important to develop an organic grouting material with strong permeability and excellent mechanical properties.

The viscosity of methyl methacrylate (MMA)-based repair material can be as low as 0.8 mPa·s, and it is even suitable for repairing microcracks with a 0.05 mm width due to its good fluidity [27,28]. Moreover, it has good chemical resistance and strong adhesion with old concrete [29]. However, the biggest disadvantage of MMA-based repair material is its high shrinkage during polymerization, which can reach about 21%; this disadvantage limits its application in repairing engineering to a great extent [30]. The main cause of volume shrinkage is the transformation of van der Waals forces between MMA monomers into covalent bonds after polymerization [31]. This results in a smaller molecular distance and thus tighter arrangement, and creates shrinkage stress that would affect the bonding effect, so it is vital to reduce its volume shrinkage [32,33]. Common methods to effectively reduce polymerization shrinkage include (1) synthesizing low-shrinkage resin, (2) adding inorganic filler and (3) adding low-shrinkage additives [34–37]. A vast number of studies on the reduction of volume shrinkage of MMA have been carried out, but there remains no guarantee that the viscosity can meet the requirements for repairing microcracks. Han [38] found that after adding perchloroethylene, the shrinkage of the MMA repair material decreased to a certain extent and then decreased significantly after the further addition of inorganic filler calcium carbonate, down to about 7%. However, the addition of calcium carbonate led to an increase in the viscosity of the resin, which prevented the calcium carbonate from being well dispersed in the system and thus reduced the bending strength of the material. Wang [39] used epoxy resin to modify the MMA material, with the results showing that the bonding strength of the repair material increased, but the viscosity also increased considerably, with continual addition of diluent being needed. Wu [40] synthesized MMA, epoxy and polyurethane prepolymer by using the interpenetrate polymer network technique. Polyurethane improved the volume stability and the flexibility but also had a significant impact on the viscosity of the system. Compared with other repair materials, MMA possesses a low viscosity as its unique advantage, so it is important to pay attention to its viscosity change while modifying it. In order to repair microcracks, in addition to maintaining good fluidity and high mechanical properties, the high shrinkage of MMA-based repair material also needs to be reduced. The addition of low-shrinkage additives can effectively reduce the shrinkage of the resin. Styrene is a nonpolar low-shrinkage additive that can form a two-phase system with resin, thereby reducing the shrinkage. The effect of reducing shrinkage is general, but its mechanical strength is relatively high. Polyvinyl acetate (PVAc) is a polar-low shrinkage additive that forms a homogeneous system with resin before curing and evenly separates phases after curing. It has an excellent shrinkage reduction effect but poor mechanical strength. PVAc + styrene low shrinkage additives have good overall performance in terms of shrinkage and mechanical strength.

In this study, in order to obtain a repair material that meets the requirements for repairing microcracks, the effects of PVAc + styrene as low-shrinkage additives on the shrinkage, viscosity, bond strength, bending strength and tensile strength of MMA-based repair material were investigated. The modification mechanism of low-shrinkage additives was also studied using SEM, FTIR and DSC.

2. Materials and Experiments

2.1. Raw Materials

Methyl methacrylate (MMA), the initiator benzoyl peroxide (BPO), the plasticizer dioctyl phthalate (DOP), the accelerator N-N-dimethylaniline (DMA) and styrene were all supplied by Yonghua Chemistry Co., Ltd. (Suzhou, China). Polyvinyl acetate (PVAc) was provided by Shanghai Meryer Chemical Technology Co., Ltd. (Shanghai, China).

2.2. Specimen Preparation

In order to ensure the normal progress of polymerization, it is necessary to remove the polymerization inhibitor which is added to MMA during the storage process. Although the content is relatively low, generally not more than 0.001%, this can also have an impact on the polymerization reaction. Therefore, the following method was used to remove the inhibitor before polymerization: MMA monomer was subjected to open distillation for 10 min in a water bath at 50 °C.

Table 1 shows the raw material ratio of the MMA-based repair material. The mass ratio of PVAc to styrene was 7:3, and PVAc was poured into styrene to dissolve. The mass ratio of MMA: BPO: DOP: DMA was 100:0.6:30:0.6. After weighing was completed, MMA, PVAc + styrene, BPO and DOP were added together into a three-mouth flask and then stirred constantly in a water bath at 80 °C. During the reaction process, in order to prevent the explosive polymerization phenomenon, it was important to be clear about the change of viscosity of the solution. If the viscosity of the system was found to be large, the three-mouth flask would be immediately removed from the water bath and cooled down.

Table 1. Raw material ratio of MMA-based repair material (wt%).

NO.	PVAc + Styrene/%	MMA	BPO	DOP	DMA	PVAc	Styrene
PS0	0	76.22	0.46	22.86	0.46	0	0
PS10	10	68.60	0.41	20.58	0.41	7.00	3.00
PS11	11	67.83	0.41	20.35	0.41	7.70	3.30
PS12	12	67.08	0.40	20.12	0.40	8.40	3.60
PS13	13	66.32	0.39	19.90	0.39	9.10	3.90
PS14	14	65.56	0.39	19.66	0.39	9.80	4.20

2.3. Testing Method

2.3.1. Shrinkage

The type of shrinkage of MMA-based repair material measured in this experiment was chemical shrinkage, which was tested using the following method: first, the volume of test tube V_1 was obtained using the mass and density of the distilled water. Then, the repair material prepared in Section 2.2 that was mixed with DMA was used to fill the test tube and was cured at room temperature. Therefore, the volume of the specimen before curing was V_1, and the mass of the specimen before curing was determined from the difference in the mass of the test tube before and after the addition of the repair material and noted as m_1. After the specimen was cured, the test tube was broken, the specimen after curing was weighed to obtain the mass m_2, and then the volume V_2 of the cured specimen was measured using the drainage method. The chemical volume shrinkage was calculated as shown in Equation (1).

$$S = \frac{\frac{m_2}{V_2} - \frac{m_1}{V_1}}{\frac{m_1}{V_1}} \times 100\% \tag{1}$$

where S is the volume shrinkage of the specimen (%); m_1 and m_2 are the mass of the specimen before and after curing (g), respectively; and V_1 and V_2 are the volume of the specimen before and after curing (mL), respectively.

2.3.2. Shrinkage Stress

The rotation rheometer of Anton Paar MCR 302 was used to test the change of internal stress of the polymer during curing under the controlled strain mode with a strain value of 1% and a frequency of 1 Hz.

2.3.3. Viscosity

The pointer type NDJ-1 rotary viscometer was used to measure the viscosity of the repair material. First, the repair material was placed in a beaker, and then the appropriate rotor and rotational speed were selected according to the viscosity of the solution. Finally, the rotor was immersed vertically in the solution, the starting switch was pressed, and the data were read after the indicator was stable.

2.3.4. Bond Strength

Ordinary Portland Cement mortar (cement:sand:water = 1:3:0.5 by weight) was used to test the bond strength. When OPC mortar was cured for 28 d under standard curing conditions (at 20 ± 2 °C and RH 95%), it was first fractured under the flexural strength test using the universal testing machine (ETM-F), and then the distance between the two fractured specimens was kept at 3 mm and pasted with tape. After the gap was filled with MMA-based repair material, the specimen was put into an oven at 40 °C for 4 h and then placed in a standard curing box; the repair model is shown in Figure 1a. The flexural strength was used as an index to measure the bond strength of the repair material and the OPC mortar. As shown in Figure 1b, if the fracture surface of the specimen appears at the bonding interface, the obtained flexural strength value is the bond strength, and if the fracture surface occurs inside the mortar specimen, the bond strength of the repair material and the OPC mortar is greater than the obtained flexural strength.

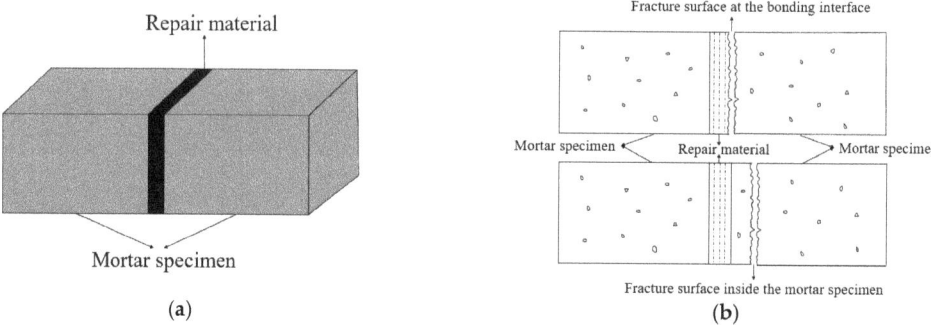

Figure 1. (a) The repair model and (b) two types of fracture surfaces.

2.3.5. Bending Strength

After accelerator DMA was added to the prepared repair material, the material was poured into a mold with dimensions of 100 mm × 15 mm × 5 mm, and finally the maximum load of bending failure of the specimen was recorded by using the strength testing machine. The bending strength of the repair material was calculated according to Equation (2):

$$R_f = \frac{3F_f L_f}{2b_f h_f^2} \qquad (2)$$

where R_f is the bending strength of the repair material (MPa); F_f is the maximum load of bending failure of the repair material (N); L_f is the distance between two points of the repair material under force (mm); and b_f and h_f are the width and height of a cross-section of repair material (mm), respectively.

2.3.6. Tensile Strength

After the repair material was poured into a mold with a size of 40 mm × 10 mm × 2 mm, tensile test was carried out on a strength testing machine. The tensile load was applied along the length of the specimen at a rate of 1 mm/min until the tensile failure of the specimen. The tensile stress–strain curve was plotted, from which the tensile strength and elongation at break of the material could be discerned. The tensile strength and elongation at break of the material were calculated according to Equations (3) and (4):

$$\sigma_t = \frac{F_t}{b_t \cdot h_t} \times 100\% \qquad (3)$$

where σ_t is the tensile strength of the material (MPa); F_t is the maximum load of tensile failure of the specimen (N); and b_t and h_t are the width and height of a cross-section of specimen fracture (mm), respectively.

$$\varepsilon_t = \frac{G_{t1} - G_{t0}}{G_{t0}} \times 100\% \qquad (4)$$

where ε_t is the elongation at a break of the specimen (%), G_{t1} is the initial marking distance of the specimen (mm), and G_{t0} is the marking distance where the specimen was fractured(mm).

2.3.7. Microscopic Characteristics and Morphology

The molecular structure of the repair material were determined with an infrared spectrometer (Nicollet Nexus 670, Shanghai, China), and the polymerization between the MMA monomer and other additives was studied by analyzing the functional groups. The glass transition temperature (T_g) of the polymer was determined with a differential scanning calorimeter (DSC 404 F1 Pegasus®, Shanghai, China). The microscopic morphology of the polymerization product was observed using SEM (JSM-6510, Shanghai, China).

3. Results and Discussion

3.1. Effect of PVAc + Styrene on Chemical Shrinkage of MMA-Based Repair Material

When no admixture was added, the chemical volume shrinkage of MMA-based concrete repair material was as high as 21%, which would definitely affect its use in repair engineering. The volume shrinkage of the material will create stress toward the interior of the material at the bonding interface, which may lead to the failure of the interface bonding or even secondary cracking [41]. Theoretically, the higher the chemical volume shrinkage of a material is, the higher its shrinkage stress, so it is necessary to add low 0hrinkage additives to reduce its shrinkage. The PVAc + styrene solution was added into MMA to prepare the repair material, and the variation law of chemical shrinkage of MMA-based repair material with different contents of PVAc + styrene was investigated, with test results being presented in Table 2.

As can be seen from the Table 2, after the addition of low-shrinkage additives, the chemical volume shrinkage and the shrinkage stress of the material decreased significantly with the increase of the proportion of additives; notably, when the proportion of PVAc + styrene was 12%, the chemical volume shrinkage was as low as 4.78%, and the shrinkage stress reduced from −5.08 N to −0.64 N, which decreased by 87.4% (negative stress values indicate shrinkage). After the addition of low-shrinkage additives, one of the reasons for the significant reduction in chemical shrinkage of the material was that polystyrene was formed in the polymerization process, and polystyrene is a nonpolar low-shrinkage additive, which can reduce the chemical shrinkage of MMA resin to a certain extent. When the proportion of PVAc + styrene exceeded 12%, the chemical shrinkage of the MMA-based repair material gradually increased with the increase in the proportion of additives. Previous study has shown that the low-shrinkage additives act only as inert materials at contents higher than the upper limit of the effective content [42]. However, compared with 22.67% of the blank specimen without additives, the chemical volume shrinkage of the repair material was still very low.

Table 2. Shrinkage results of MMA-based materials with different PVAc + styrene contents.

NO.	Mass/g	Before Curing Volume /mL	Density /g·mL^{-1}	Mass/g	After Curing Volume /mL	Density /g·mL^{-1}	Volume Shrinkage/%	Shrinkage Stress/N
PS0	3.75	4.30	0.87	3.68	3.44	1.07	22.67	−5.08 *
PS10	4.17	4.10	1.02	3.88	3.55	1.09	7.46	−2.23
PS11	4.20	4.05	1.04	4.16	3.75	1.11	6.97	−0.78
PS12	4.19	4.02	1.04	4.03	3.69	1.09	4.78	−0.64
PS13	4.26	4.29	0.99	4.19	3.83	1.09	10.17	−0.71
PS14	4.18	4.39	0.95	3.98	3.65	1.09	14.51	−2.95

* Negative stress values indicate shrinkage.

3.2. Effect of PVAc + Styrene on the Polymerization Process of MMA-Based Repair Material

Viscosity is used to characterize the rheological property of the polymer, which varies with time [43]. A low viscosity of the polymer means that it has better fluidity and thus can penetrate into microcracks more easily. One of the most prominent advantages of MMA-based repair material compared with other repair materials is its low viscosity, which can repair microcracks smaller than 0.5 mm in width, so it is vital to test the change rule of the viscosity of MMA-based repair material in modification experiments. The effect of the reaction time on the viscosity of the MMA-based repair material was investigated by testing the viscosity of the repair materials with 0% and 12% PVAc + styrene, and the change in shrinkage stress during the polymerization was also measured to determining the reason for its reduction, with the results being shown in Figure 2.

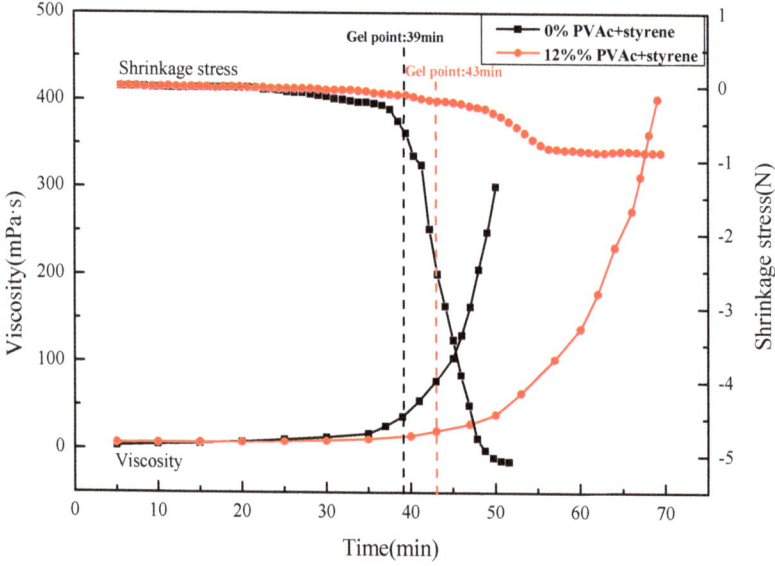

Figure 2. The curing process of MMA-based repair materials with different dosages of PVAc + styrene.

The polymerization process of MMA, as shown in Figure 2, is similar to that reported in the previous research results, which can be basically divided into two stages, and the viscosity increases approximately exponentially with time [44,45]. Before gelation, the initiator decomposes into free radicals, so the viscosity almost remains constant. Then after gelation, with the increase in the number of free radicals and chain activation centers, the main chain of PMMA increases, and the viscosity of the system also changes significantly with reaction time. It should be noted that as the reaction proceeds, the viscosity increases sharply, making it difficult to remove the polymerization heat from the system, so it is necessary to control the reaction time to prevent explosion at this stage. The inflection point of the curve is considered to be the gel point. Since the movement of the macromolecules is relatively free before gelation, the shrinkage stress can be effectively released, while after gelation, the spatial network structure is formed, the movement of the molecular chains is difficult, and so the shrinkage stress gradually increases. Therefore, it can be seen that the shrinkage stress of the repair materials without and with 12% PVAc + styrene increases sharply after 39 min and 43 min of the gel point until reaching the maximum value, respectively.

In addition, by comparing the viscosity of MMA-based repair materials without PVAc + styrene and with 12% PVAc + styrene at the same time, we found that the gel point of the polymerization process was delayed after adding low-shrinkage additives. This is because with the addition of additives, the free radicals decomposed by the initiator activate not only the MMA molecules but also the styrene molecules. Therefore, the growth of the main chain of PMMA slows down, the rate of forming network structure decreases correspondingly, and the fluidity of the main chain increases, thus reducing the shrinkage stress. Therefore, the delay in the gel point retards the development of the stress and thus reduces the shrinkage of the material, which is similar to results reported elsewhere [46].

3.3. The Shrinkage Reduction Mechanism of PVAc + Styrene on MMA-Based Repair Material

In order to investigate the effect of PVAc + styrene on the structure of MMA-based repair material, MMA repair materials without and with 12% PVAc + styrene were tested using FTIR, DSC and SEM.

Figure 3 shows the FTIR spectra for the MMA-based repair materials with the ratios of PS0 and PS12 in the range of 500–4000 cm^{-1}, with m1 being the PS0 repair material and m2 being the PS12 repair material. It can be noticed that there was no characteristic peak of C=C in either curve, indicating that during the reaction, both MMA and styrene underwent polymerization. In the m1 curve, the presence of the bands representing $-CH_3$, $-CH_2-$, C=O and C-O demonstrated the formation of polymethyl methacrylate. In the m2 curve, in addition to all the absorption bands of m1, the bands at 3021 cm^{-1} and 704 cm^{-1} were associated with the vibration of C-H in the benzene ring, the bands at 1484 cm^{-1} and 1435 cm^{-1} were attributed to the stretching vibration of C-C in the benzene ring, and these bands represented the formation of polystyrene.

Subsequently, for specifying whether the polymerization between MMA and styrene occurred, DSC analysis was carried out. MMA-based repair materials with and without 12% PVAc + styrene were ground into powder for DSC testing, and the obtained experimental results were shown in Figure 4.

As can be seen from the Figure 4, the glass transition temperature (T_g) of the MMA-based repair material was about 111.1 °C without the low-shrinkage additives, while the T_g of the material decreased to 60.8 °C after the low-shrinkage additives were incorporated. This is mainly because after the addition of the low-shrinkage additives, the free radicals decomposed by the initiator activated not only the MMA molecules but also styrene molecules so that in the heating process, MMA and styrene in the system underwent copolymerization, which reduced the glass transition temperature.

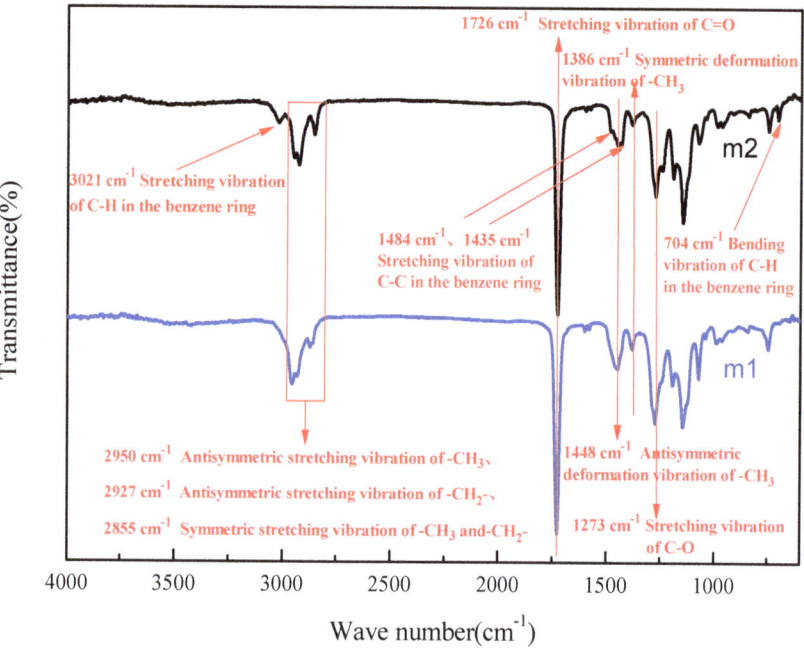

Figure 3. FTIR spectra of the polymerization products of PS0(m1) and PS12(m2).

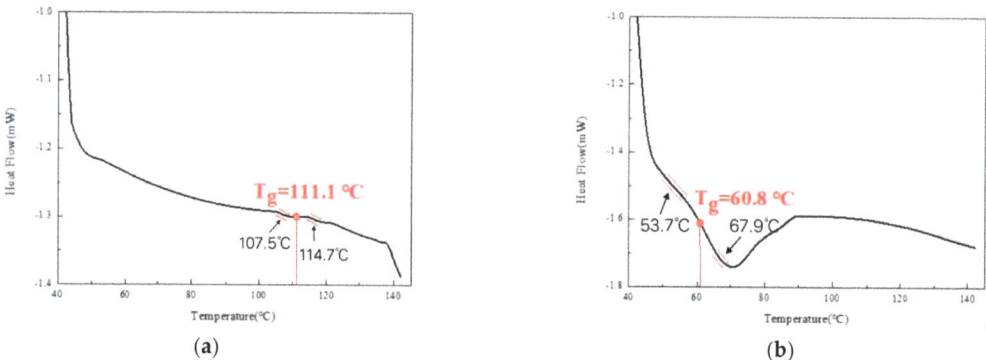

Figure 4. DSC curve of repair material (**a**) without and (**b**) with 12% PVAc + styrene.

The fractured surfaces of polymerization products with the ratios of PS0 and PS12 were also observed using SEM. In Figure 5a of, it can be seen from the specimen without additives that the overall morphology was a single-phase structure, dominated by the dense MMA continuous phase. As seen in Figure 5b,c with 12% PVAc + styrene, the fractured surface of the specimen showed a two-phase structure clearly, with the PVAc phase dispersed in the MMA phase in the form of particles. There were microvoids around the particles, which could compensate for the volume shrinkage to some extent. In addition, in Figure 5c, it can be seen that the polystyrene was unevenly dispersed in the MMA phase in the form of microbeads, which counteracted the volume shrinkage of the resin with its own sufficient expansion.

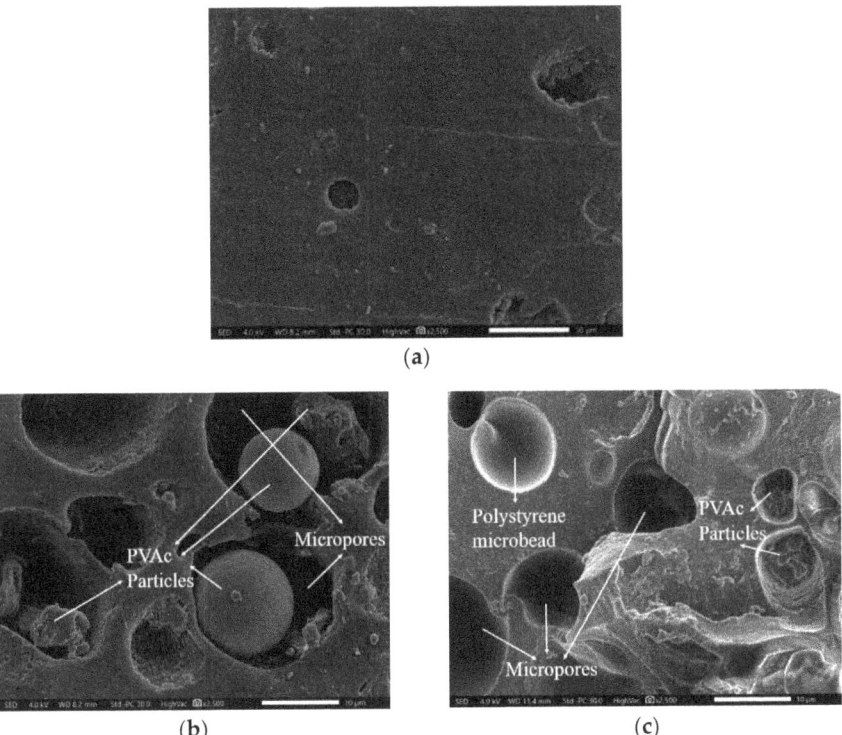

Figure 5. SEM micrographs of the polymerization products with different contents of PVAc + styrene. (**a**) No PVAc + styrene; (**b**,**c**) 12% PVAc + styrene.

As seen in Table 2, the shrinkage of MMA-based repair materials begins to decrease when the PVAc + styrene is greater than 12%. The reasons for the decrease were investigated, and the SEM of the polymerization products of the MMA-based repair materials of PS12 and PS14 is shown in Figure 6.

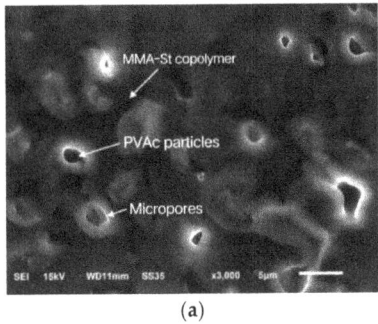
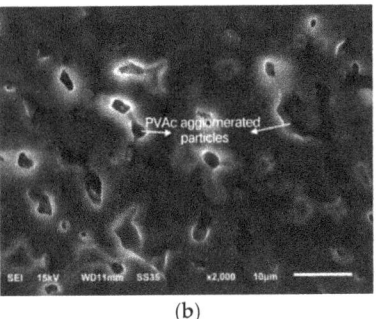

Figure 6. SEM micrographs of the polymerization products with different contents of PVAc + styrene. (**a**) 12% PVAc + styrene; (**b**) 14% PVAc + styrene.

As can be seen from Figure 6a, when the dosage of PVAc + styrene was 12%, the PVAc particles were spherically dispersed in the repair material, with a size of about 1–3 μm, and there were micropores around them with a size less than 1 μm, which considerably reduced the shrinkage. When the dosage of PVAc + styrene increased to 14%, as shown in

Figure 6b, the size of the PVAc particles became larger due to the agglomeration, showing lumpy and long stripes, and the size of some particles was even larger than 5 µm, which would affect its uniformity in the repair material and reduce the number and volume of micropores, making the effect of shrinkage reduction worse.

From the above analysis of the copolymerization products, including FTIR spectra, DSC curves and SEM images, combined with the analysis on the shrinkage characteristics of unsaturated resins [47–49], the shrinkage reduction mechanism of this system can be deduced as follows:

1. Phase separation stage: PVAc is dissolved in styrene solution as sphere-like particles, and each particle occupies a certain volume. However, styrene is a nonpolar solvent, while PVAc is a polar polymer, which eventually leads to a decrease in the compatibility of PVAc with styrene, resulting in phase separation, which is shown schematically in Figure 7a.

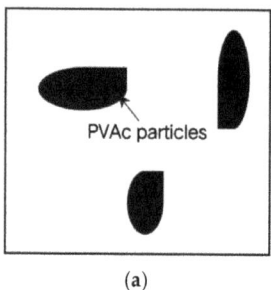 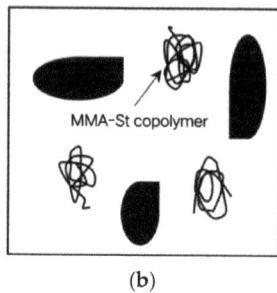 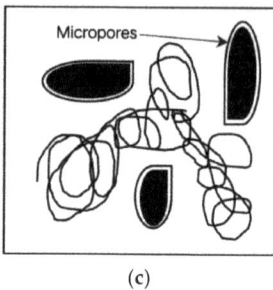

(a) (b) (c)

Figure 7. Mechanism of PVAc + styrene for MMA repair material. (**a**) Stage of phase separation; (**b**) Stage of macroscopic gel phase; (**c**) Stage of creating microspores and expanded.

2. Macroscopic gel phase stage: In the process of polymerization, when the temperature rises, the initiator BPO begins to decompose into free radicals, which undergo addition reactions with MMA or styrene. During this period, the polymerization reactions between MMA and MMA, between styrene and styrene and between MMA and styrene take place, forming a continuous phase of MMA, a dispersed phase of PVAc and a random copolymer of MMA-St within the system, which gradually produces a macroscopic gel as the degree of polymerization increases. The final structure of the copolymer surrounded by PVAc particles is shown in Figure 7b. In the newly formed structure, the benzene ring linked to the main chain of MMA produces repulsion and hindrance, forming a steric effect, which reduces the volume shrinkage to a certain extent.

3. Microspore creation and expansion stage: As shown in Figure 7c, when the reaction rate is about to reach its maximum, PVAc plays a role in accumulating styrene, as the heat formed by the polymerization reaction allows PVAc and unreacted styrene monomer to expand, compensating for the volume shrinkage. As the temperature of the system decreases, MMA phase and PVAc phase shrink at the same time. When the temperature is higher than $T_{g,PMMA}$, the coefficients of thermal expansion of these two phases are basically the same. When the temperature is reduced to higher than $T_{g,PVAc}$ and lower than $T_{g,PMMA}$, the contraction effect of the PVAc phase is greater than that of the MMA phase, resulting in the formation of micropores at the interface between the two phases, and the micropores become larger and larger until the temperature is gradually reduced to $T_{g,PVAc}$, ($T_{g,PVAc}$ is 30 °C, $T_{g,PMMA}$ is 105 °C). During the formation of micropores, the internal stress is partially released, so the shrinkage stress of the material is reduced after the addition of low-shrinkage additives.

3.4. Effect of PVAc + Styrene on the Mechanical Properties of MMA-Based Repair Material

Based on the results in Sections 3.1–3.3, it can be concluded that PVAc + styrene can improve the shrinkage of MMA-based repair material to a certain extent, but its effect on the mechanical properties of repair material still needs to be investigated. The interface of organic–inorganic composites is the main factor affecting their overall mechanical properties, so it is important to study the interfacial bonding property. The results of the bond strength of repair materials with different PVAc + styrene ratios at different ages are shown in the Table 3.

Table 3. Effect of PVAc + styrene contents on bond strength of MMA-based repair materials.

No.	3 d		7 d		28 d	
	Bond Strength/MPa	Fracture Location	Bond Strength/MPa	Fracture Location	Bond Strength/MPa	Fracture Location
PS0	>4.62 *	M *	>5.32	M	>5.51	M
PS10	2.60	I *	3.90	I	>4.21	M
PS11	3.84	I	>4.07	M	>4.39	M
PS12	2.93	I	>4.04	M	>4.10	M
PS13	2.83	I	>3.97	M	>4.13	M
PS14	2.63	I	3.83	I	>4.03	M

* "M" represents the fracture surface occurring inside the mortar specimen. "I" represents the fracture surface appearing at the bonding interface. ">" represents a bond strength greater than the obtained flexural strength.

As can be seen from the Table 3, the variation of bond strength at different ages with different proportions is basically the same. At 3 d, the bond strength of the repair material with low-shrinkage additives was relatively low, and all specimens fractured at the bonding interface. At 7 d, the bond strength was significantly higher than that at 3 d, and some specimens had fractures inside the OPC mortar. The specimens at 28 d all had fractures inside the OPC mortar, implying that the bond strength was greater than the strength value obtained, which was highly beneficial for the actual repair engineering. The two types of fracture surfaces of bonding specimens are shown in Figure 8. Figure 8a shows that the fracture surface appeared at the bonding interface between the repair material and the OPC mortar, indicating that the bonding surface was the weak link of the specimen. Figure 8b shows that the fracture surface occurred inside the OPC mortar, demonstrating a good interfacial bonding performance.

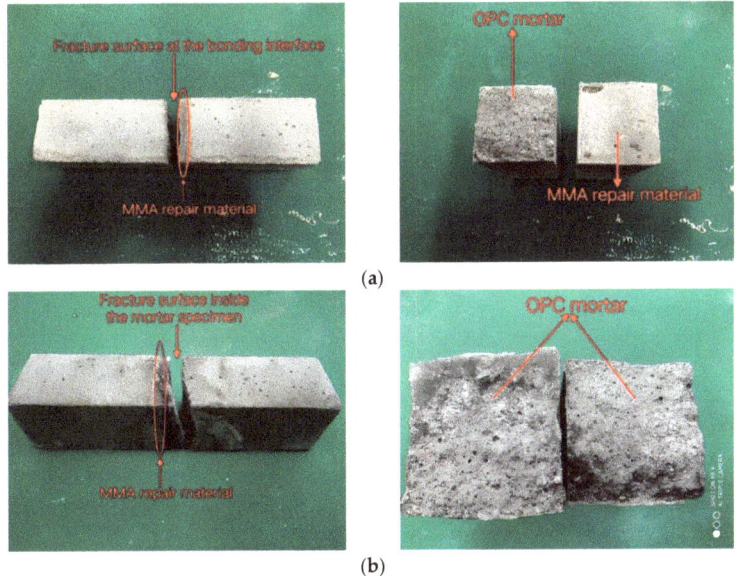

Figure 8. Two types of fracture surfaces: (**a**) fracture surface at the bonding interface and (**b**) fracture surface inside the OPC mortar.

Comparing the bond strength of the repair material without and with PVAc + styrene, we can surmise that after adding PVAc + styrene, the bond strength of repair material decreased, especially in the early stage. This may be because the addition of PVAc + styrene causes a significant decrease in shrinkage stress, leading to the reduction of local void phenomenon at the bonding interface, as shown in Figure 9, and the improvement of bond strength. However, at the same time, micropores generated inside the material, as shown in SEM micrographs, reduce the bond strength, and the degree of reduction is greater than the degree of improvement, so the overall bonding performance of the material becomes worse.

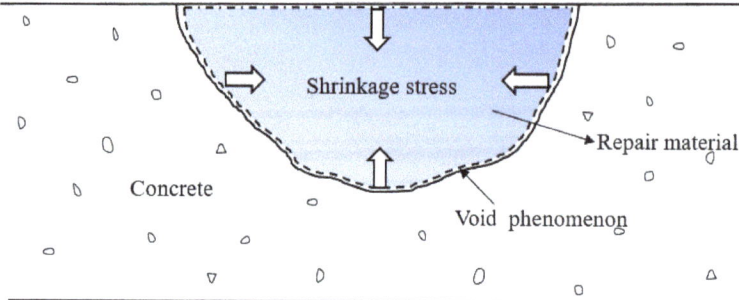

Figure 9. Shrinkage interface between the repair material and concrete.

As a material for repairing cracks, it should be able to resist the bending moment without fracture within a certain range. To determine this fracture resistance, this experiment tested the effect of different proportions of PVAc + styrene on the bending strength of MMA-based repair material; the results obtained are shown in Table 4.

Table 4. Bending strength of the repair materials.

No.		PS0	PS10	PS11	PS12	PS13	PS14
Bending	3d	19.69	21.73	22.97	20.74	18.78	24.28
strength	7d	25.93	31.38	30.25	26.25	23.79	31.11
/MPa	28d	27.52	31.48	31.56	28.04	26.17	32.81

It can be seen from the Table 4 that with the extension of the curing time, the bending strength of the material gradually increased and the difference between the bending strength at 7 d and 28 d was not significant. After the addition of additives, the change trend of the bending strength became complex, but most of the bending strengths increased to a certain extent. When the proportion of PVAc + styrene was 14%, the bending strength of the material could reach 32.81 MPa. This is mainly because after the admixture of additives, during the polymerization reaction, the free radicals decomposed by the initiator activate not only the MMA molecules but also the styrene molecules, so the growth rate of PMMA main chain slows down. At the same time, during the curing process, the steric hindrance generated would reduce the reaction rate and the heat release, so as to avoid the internal scorching of the repair material due to the large amount of exotherm, which has little impact on the bending strength. Therefore, the incorporation of low-shrinkage additives results in a substantial increase in the bending strength of the repair material.

Tensile strength is one of the most important mechanical properties and reflects the resistance of the material to the lateral forces after repair engineering. Two repair materials, PS0 and PS12, were used to make tensile specimens and conduct the experiments, and the tensile stress–strain curves were plotted as shown in Figure 10.

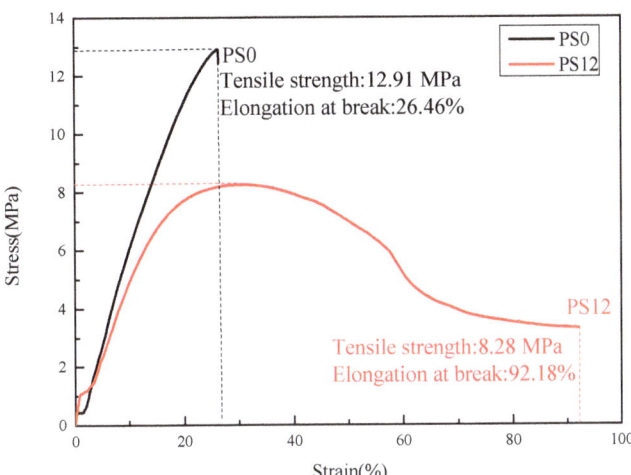

Figure 10. Tensile stress–strain curve of the MMA-based repair material.

From the curves, it can be discerned that the specimen without PVAc + styrene was pulled out with a tensile load up to 258.26 N, the corresponding elongation at break was 26.46%, the tensile strength was 12.91 MPa, and the elastic modulus was 70.79 MPa. After the addition of 12% PVAc + styrene, the material broke at a tensile load of 165.57 N, with an elongation at break of 92.18%, a tensile strength of 8.28 MPa and an elastic modulus of 54.35 MPa. As seen in Figure 10, the stress–strain curves of the repair materials were basically smooth, which is consistent with the characteristics of plastic materials. When PVAc + styrene was not incorporated, the deformation of the material after yielding was small. At this time, the elongation at the break of the repair material was small, and the tensile strength and elastic modulus were large, demonstrating hard and strong characteristics. After the incorporation of 12% PVAc + styrene, the deformation of the material after yielding was greater, the elongation at the break increased significantly, with improved fracture toughness of the material.

It can be seen that the addition of 12% PVAc + styrene reduced the elastic modulus of MMA-based repair material. Generally, the lower elastic modulus corresponded to the smaller shrinkage stress. This occurred because the lower elastic modulus can make the stress dissipate better in the polymerization process, so the shrinkage stress was 87.4% lower than that of the material without low-shrinkage additives, as shown in Table 2. Because the generated two-phase system had poor compatibility and was not conducive to stress transmission, the tensile strength of the repair material added with low-shrinkage additives decreased, and the existence of micropores and steric hindrance also affected it. Although the tensile strength of the material decreased, its elongation at break increased significantly, indicating that it would not cause brittle fracture immediately when subjected to tension.

From the measurements of shrinkage, viscosity and mechanical properties, it can be concluded that PVAc and styrene as a low-shrinkage additives had a considerable effect on the overall performance of the MMA-based repair material. Based on the results of this study, it was found that the addition of low-shrinkage additives had a favorable effect on the shrinkage and shrinkage stress of the repair material but a more variable effect on the mechanical properties. This is because the two-phase structure and the formation of micropores compensate for the volume shrinkage and reduce the shrinkage stress, but the presence of micropores also has an adverse effect on some mechanical properties of the material. It is possible to obtain MMA-based repair material with specific requirements by varying the content of low-shrinkage additives to repair concrete microcracks.

4. Conclusions

The main conclusions are as follows:

1. The addition of the low-shrinkage additives of PVAc and styrene significantly reduced the volume shrinkage and shrinkage stress of the MMA-based repair material and delayed the polymerization reaction to some extent. When the content of PVAc + styrene was 12%, the volume shrinkage of the repair material was the lowest, only 4.78%, and the shrinkage stress decreased by 87.4%.

2. During the polymerization process, styrene and MMA undergo copolymerization, which reduces the volume shrinkage to a certain extent through a steric hindrance effect. The polystyrene generated during the polymerization counteracts the curing shrinkage mainly by thermal expansion. The formation of micropores at the interface between the continuous phase of MMA and the dispersed phase of PVAc also offsets part of the volume shrinkage.

3. After the addition of low-shrinkage additives, the bond strength of the MMA-based repair material was relatively low after 3 d, and the fracture surfaces all appeared at the bonding interface between the repair material and the OPC mortar. At 28 d, the specimens all had fractures inside the OPC mortar in testing, showing a good bonding performance.

4. PVAc + styrene improved the bending strength of most of the repair materials in the ratios investigated in this study; when the proportion was 12%, the 28 d bending strength of the material reached 28.04 MPa, and the elongation at break and the tensile strength of the MMA-based repair material were 92.18% and 8.28 MPa, respectively.

Author Contributions: Conceptualization, Z.G. and L.X.; methodology, Z.G. and L.X.; data curation, Z.G., S.L., L.Y., Z.Z. and Y.W.; testing and providing help in calculation, Z.G., S.L., L.Y., Z.Z. and Y.W.; writing—original draft preparation, Z.G.; writing—review and editing, Z.G. and L.X.; All authors have read and agreed to the published version of the manuscript.

Funding: The work was supported by the Priority Academic Program Development of Jiangsu Higher Education Institutions (PAPD).

Institutional Review Board Statement: Not applicable.

Informed Consent Statement: Not applicable.

Data Availability Statement: All data in this article are listed in this paper. The data presented in this study are available on request from the corresponding author.

Acknowledgments: The work was supported by the Priority Academic Program Development of Jiangsu Higher Education Institutions (PAPD). In addition, the authors appreciate the efforts of the anonymous reviewer who improved the quality of this paper.

Conflicts of Interest: The authors declare no conflict of interest.

References

1. Fu, C.-Q.; Fang, D.-M.; Ye, H.-L.; Huang, L.; Wang, J.-D. Bond degradation of non-uniformly corroded steel rebars in concrete. *Eng. Struct.* **2021**, *226*, 111392. [CrossRef]
2. Yu, H.-F.; Lu, L.-J.; Qiao, P.-Z. Localization and size quantification of surface crack of concrete based on rayleigh wave attenuation model. *Constr. Build. Mater.* **2021**, *280*, 122437. [CrossRef]
3. Dumounlin, C.; Deraemaeker, A. Real-time fast ultrasonic monitoring of concrete cracking using embedded piezoelectric transducers. *Smart Mater. Struct.* **2017**, *26*, 104006. [CrossRef]
4. Ghaffary, A.; Moustafa, M.-A. Synthesis of repair materials and methods for reinforced concrete and prestressed bridge girders. *Materials* **2020**, *13*, 4079. [CrossRef]
5. Song, X.-M.; Song, X.-F.; Liu, H.; Huang, H.-L.; Auvarovna, K.-G.; Ugli, N.-A.-D.; Huang, Y.; Hu, J.; Wei, J.-X.; Yu, Q.-J. Cement-based repair materials and the interface with concrete substrates: Characterization, evaluation and improvement. *Polymers* **2022**, *14*, 1485. [CrossRef]
6. Liu, Y.-S.; Jia, M.-J.; Song, C.Z.; Lu, S.; Wang, H.; Zhang, G.-H.; Yang, Y.-S. Enhancing ultra-early strength of sulphoaluminate cement-based materials by incorporating graphene oxide. *Nanotechnol. Rev.* **2020**, *9*, 17–27. [CrossRef]
7. Garcia, C.-J.-L.; Pedrosa, F.; Carballosa, P.; Revuelta, D. Evaluation of the sealing effectiveness of expansive cement grouts through a novel water penetration test. *Constr. Build. Mater.* **2020**, *251*, 118974. [CrossRef]

8. Liu, F.; Pan, B.-F.; Zhou, C.-J. Experimental study on a novel modified magnesium phosphate cement mortar used for rapid repair of portland cement concrete pavement in seasonally frozen areas. *J. Mater. Civ. Eng.* **2022**, *34*, 04021483. [CrossRef]
9. Safan, M.-A.; Etman, Z.-A.; Konswa, A. Evaluation of polyurethane resin injection for concrete leak repair. *Case Stud. Constr. Mater.* **2019**, *11*, e00307. [CrossRef]
10. Haufe, M.; Bakalli, M.; Sahin, M.; Helm, C.; Achenbach, R. Newly developed acrylate injection resin gels for structural sealing and active corrosion protection of embedded steel reinforcement. *Mater. Corros.* **2020**, *71*, 824–833. [CrossRef]
11. Fang, Y.-F.; Ma, B.-A.; Wei, K.; Wang, L.; Kang, X.-X.; Chang, X.-G. Orthogonal experimental analysis of the material ratio and preparation technology of single-component epoxy resin for asphalt pavement crack repair. *Constr. Build. Mater.* **2021**, *288*, 123074. [CrossRef]
12. Kim, T.-K.; Park, J.-S. Comparative analysis of domestic and international test guidelines for various concrete repair materials. *Materials* **2022**, *15*, 3267. [CrossRef] [PubMed]
13. Shi, C.; Wang, P.; Ma, C.-Y.; Zou, X.-Y.; Yang, L. Effects of SAE and SBR on properties of rapid hardening repair mortar. *J. Build. Eng.* **2021**, *35*, 102000. [CrossRef]
14. Lakhiar, M.-T.; Bai, Y.; Wong, L.-S.; Paul, S.-C.; Anggraini, V.; Kong, S.-Y. Mechanical and durability properties of epoxy mortar incorporating coal bottom ash as filler. *Constr. Build. Mater.* **2022**, *315*, 125677. [CrossRef]
15. Drochytka, R.; Hodul, J.; Meszarosova, L.; Jakubik, A. Chemically resistant polymeric jointing grout with environmental impact. *Constr. Build. Mater.* **2021**, *292*, 123454. [CrossRef]
16. Zhang, S.; Qiao, W.-G.; Chen, P.-C.; Xi, K. Rheological and mechanical properties of microfine-cement-based grouts mixed with microfine fly ash,colloidal nanosilica and superplasticizer. *Constr. Build. Mater.* **2019**, *212*, 10–18. [CrossRef]
17. Li, X.-L.; Hao, M.-M.; Zhong, Y.-H.; Zhang, B.; Wang, F.-M.; Wang, L.B. Experimental study on the diffusion characteristics of polyurethane grout in a fracture. *Constr. Build. Mater.* **2021**, *273*, 121711. [CrossRef]
18. Abolfazli, M.; Bazli, M.; Heydari, H.; Fahimifar, A. Investigating the effects of cement and polymer grouting on the shear behavior of rock joints. *Polymers* **2022**, *14*, 1229. [CrossRef]
19. Anagnostapoulos, C.-A.; Papaliangas, T.; Manolopoulou, S.; Dimopoulous, T. Physical and mechanical properties of chemically grouted sand. *Tunn. Undergr. Space Technol.* **2011**, *26*, 718–724. [CrossRef]
20. Jin, L.-C.; Sui, W.-H.; Xiong, J.-L. Experimental investigation on chemical grouting in a permeated fracture replica with different roughness. *Appl. Sci.* **2019**, *9*, 2762. [CrossRef]
21. Wang, W.-Z.; Zhao, W.-Q.; Zhang, J.-J.; Zhou, J.-H. Epoxy-based grouting materials with super-low viscosities and improved toughness. *Constr. Build. Mater.* **2021**, *267*, 121104. [CrossRef]
22. Yin, H.; Sun, D.-W.; Li, B.; Liu, Y.-T.; Ran, Q.-P.; Liu, J.-P. DSC and curing kinetics study of epoxy grouting diluted with furfural-acetone slurry. In Proceedings of the Global Conference on Polymer and Composite Materials (PCM), Hangzhou, China, 20–23 May 2016. [CrossRef]
23. Kumar, S.; Krishnan, S.; Samal, S.-K.; Mohanty, S.; Nayak, S.-K. Toughening of petroleum based (DGEBA) epoxy resins with various renewable resources based flexible chains for high performance applications: A review. *Ind. Eng. Chem. Res.* **2018**, *57*, 2711–2726. [CrossRef]
24. Hao, M.-M.; Li, X.-L.; Zhong, Y.-H.; Zhang, B.; Wang, F.-M. Experimental study of polyurethane grout diffusion in a water-bearing fracture. *J. Mater. Civ. Eng.* **2021**, *33*, 04020485. [CrossRef]
25. Yuan, J.-Q.; Chen, W.-Z.; Tan, X.-J.; Yang, D.-S.; Zhang, Q.-Y. New method to evaluate antiwashout performance of grout for preventing water-inrush disasters. *Int. J. Geomech.* **2020**, *20*, 06019021. [CrossRef]
26. Wang, X.-M.; Li, Q. A new method for preparing low viscosity and high solid content waterborne polyurethane-phase inversion research. *Prog. Org. Coat.* **2019**, *131*, 285–290. [CrossRef]
27. Genedy, M.; Matteo, E.-N.; Stenko, M.; Stormont, J.-C.; Taha, M.-R. Nanomodified methyl methacrylate polymer for sealing of microscale defects in wellbore systems. *J. Mater. Civ. Eng.* **2019**, *31*, 04019118. [CrossRef]
28. Geng, F.; Gao, P.-W.; Xu, S.-Y.; Liu, B.; Lu, Z.-P.; Lou, S.-N. Study on repairing material made of high performance crylic acid species for concrete pavement cracks. *J. Nanjing Univ. Aeronaut. Astronaut.* **2013**, *45*, 255–259. [CrossRef]
29. Tomic, N.-Z.; Marinkovic, A.-D.; Veljovic, D.; Trifkovic, K.; Levic, S.; Radojevic, V.; Heinemann, R.-J. A new approach to compatibilization study of EVA/PMMA polymer blend used as an opticalfibers adhesive: Mechanical, morphological and thermal properties. *Int. J. Adhes. Adhes* **2018**, *81*, 11–20. [CrossRef]
30. Haas, S.-S.; Brauer, G.-M.; Dickson, G. A characterization of polymethylmethacrylate bone cement. *J. Bone. Joint. Surg. Am.* **1975**, *57*, 380–391. [CrossRef]
31. Kim, L.-U.; Kim, J.-W.; Kim, C.-K. Effects of molecular structure of the resins on the volumetric shrinkage and the mechanical strength of dental restorative composites. *Biomacromolecules* **2006**, *7*, 2680–2687. [CrossRef]
32. Qin, R.-Y.; Hao, H.-L.; Rousakis, T.; Lau, D. Effect of shrinkage reducing admixture on new-to-old concrete interface. *Compos. Part B-Eng.* **2019**, *167*, 346–355. [CrossRef]
33. Par, M.; Burrer, P.; Prskalo, K.; Schmid, S.; Schubiger, A.-L.; Marovic, D.; Tarle, Z.; Attin, T.; Taubock, T.-T. Polymerization kinetics and development of polymerization shrinkage stress in rapid high-intensity light-curing. *Polymers* **2022**, *14*, 3296. [CrossRef]
34. Hong, C.-M.; Wang, X.-J.; Kong, P.; Pan, Z.-G.; Wang, X.-Y. Effect of succinic acid on the shrinkage of unsaturated polyester resin. *J. Appl. Polym. Sci.* **2015**, *132*, 41276. [CrossRef]

35. Park, H.-S.; Sim, H.-J.; Oh, H.-K.; Lee, G.-H.; Kang, M.-A.; Lyu, M.-Y. Experimental and computational study on the mold shrinkage of PPS resin in injection molded specimen. *Elastom. Compos.* **2020**, *55*, 120–127. [CrossRef]
36. Luo, S.-Z.; Liu, F.; He, J.-W. Preparation of low shrinkage stress dental composite with synthesized dimethacrylate oligomers. *J. Mech. Behav. Biomed. Mater.* **2019**, *94*, 222–228. [CrossRef]
37. Yan, Y.; Wang, H.; Wang, X.-J.; Zhu, W.-W. Effect of 2,2-dimethyl malonate on the volume shrinkage control of the $CaCO_3$/unsaturated polyester resin composites. *J. Appl. Pplym. Sci.* **2012**, *124*, 4606–4611. [CrossRef]
38. Han, J.; Xu, L.-L.; Feng, T.; Shi, X.; Zhang, P. Effect of PCE on properties of MMA-based repair material for concrete. *Materials* **2021**, *14*, 859. [CrossRef]
39. Wang, X. Study on grouting material for mortar layer crack of CRTSIslab ballastless track. *New Chem. Mater.* **2018**, *46*, 250–252, 256. [CrossRef]
40. Wu, H.; Zhu, M.-J.; Liu, Z.; Yin, J. Developing a polymer-based crack repairing material using interpenetrate polymer network (IPN) technology. *Constr. Build. Mater.* **2018**, *84*, 192–200. [CrossRef]
41. Shan, P.-K.; Stansbury, J.-W. Photopolymerization shrinkage-stress reduction in polymer-based dental restoratives by surface modification of fillers. *Dent. Mater.* **2021**, *37*, 578–587. [CrossRef]
42. Assmuth, V.; Meschut, G. Effect of low-profile additives on the durability of adhesively bonded sheet moulding compound (SMC). *Int. J. Adhes. Adhes.* **2022**, *112*, 103036. [CrossRef]
43. Su, Z.-H.; Wang, Z.-Q.; Zhang, D.; Wei, T. Study on rheological behavior and surface properties of epoxy resin chemical grouting material considering time variation. *Materials* **2019**, *12*, 3277. [CrossRef] [PubMed]
44. Funehag, J.; Thorn, J. Radial penetration of cementitious grout-laboratory verification of grout spread in a fracture model. *Tunn. Undergr. Space Technol.* **2018**, *72*, 228–232. [CrossRef]
45. Hao, M.-M.; Li, X.-L.; Wang, X.-L.; Zhong, Y.-H.; Zhang, B.; Wang, F.-M.; Zhang, Y.-L. Experimental study on viscosity characteristics of expanding polymer grout. *J. Wuhan Univ. Technol.* **2021**, *36*, 297–302. [CrossRef]
46. Sun, G.-Q.; Wu, X.-Y.; Liu, R. A comprehensive investigation of acrylates photopolymerization shrinkage stress from micro and macro perspectives by real time MIR-photo-rheology. *Prog. Org. Coat.* **2021**, *155*, 106229. [CrossRef]
47. Zare, L.; Beheshty, M.-H.; Vafayan, M. The evaluation of low profile additives on the shrinkage control of unsaturated polyester restin cured at room temperature. *Polym. Polym. Compos.* **2012**, *20*, 289–297. [CrossRef]
48. Zong, D.-D.; Zhang, Y.; Chen, J.-R.; Liu, Y.-J. Influencing factors in low temperature curing shrinkage of unsaturated polyester resin. *J. Mater. Sci. Eng.* **2021**, *39*, 77–81+100. [CrossRef]
49. Zhang, Z.; Zhu, S. Microvoids in unsaturated polyester resins containing poly(vinyl acetate) and composites with calcium carbonate and glass fibers. *Polymer* **2000**, *41*, 3861–3870. [CrossRef]

Disclaimer/Publisher's Note: The statements, opinions and data contained in all publications are solely those of the individual author(s) and contributor(s) and not of MDPI and/or the editor(s). MDPI and/or the editor(s) disclaim responsibility for any injury to people or property resulting from any ideas, methods, instructions or products referred to in the content.

Article

Effect of Mineral Admixtures on the Mechanical and Shrinkage Performance of MgO Concrete

Xuan Zhou [1], Zhongyang Mao [1,2], Penghui Luo [1,2] and Min Deng [1,2,*]

[1] College of Materials Science and Engineering, Nanjing Tech University, Nanjing 211800, China; 202061203292@njtech.edu.cn (X.Z.); mzy@njtech.edu.cn (Z.M.); 202162103025@njtech.edu.cn (P.L.)

[2] State Key Laboratory of Materials-Oriented Chemical Engineering, Nanjing 211800, China

* Correspondence: dengmin@njtech.edu.cn; Tel.: +86-136-0518-4865

Abstract: Shrinkage deformation of concrete has been one of the difficulties in the process of concrete performance research. Cracking of concrete caused by self-shrinkage and temperature-drop shrinkage has become a common problem in the concrete world, and cracking leads to a decrease in the durability of concrete and even a safety hazard. Mineral admixtures, such as fly ash and mineral powder, are widely used to improve the temperature drop shrinkage of mass concrete; fly ash can reduce the temperature rise of concrete while also reducing the self-shrinkage of concrete, there are different results on the effect of mineral powder on the self-shrinkage of concrete, but the admixture of fly ash will reduce the strength of concrete, and mineral admixtures have an inhibitory effect on the shrinkage compensation effect of MgO expander(MEA). The paper investigates the effect of mineral admixtures on the mechanical and deformation properties of C50 mass concrete with a MgO expander(MEA), aiming to determine the proportion of C50 mass concrete with good anti-cracking properties under working conditions. The experiments investigated the effect of fly ash admixture, mineral powder admixture and MgO expander admixture on the compressive strength and deformation of concrete under simulated working conditions of variable temperature and analyzed the effect of hydration of magnesite in MgO expander and pore structure of cement paste on deformation. The following main conclusions were obtained: 1. When the concrete compounded with mineral admixture was cured under variable temperature conditions, the compounded 30% fly ash and mineral powder decreased by 4.3%, 6.0% and 8.4% at 7d age, and the compounded 40% fly ash and mineral powder decreased by 3.4%, 2.8% and 2.3% at 7d age, respectively. The incorporation of MEA reduced the early compressive strength of concrete; when the total amount of compounding remained unchanged, the early compressive strength of concrete was gradually smaller as the proportion of compounding decreased. 2. The results of concrete deformation showed that when the temperature rose, the concrete expanded rapidly, and when the temperature dropped, the concrete also showed a certain shrinkage, and the deformation of concrete basically reached stability at 18d. 3. The compounding of 30% fly ash and mineral powder As the compounding ratio decreases, the deformation of concrete increases, and the 28d deformation of concrete with a compounding ratio of 2:1 is 280×10^{-6}, while the final stable deformation of concrete with a compounding ratio of 2:1 in compounding 40% fly ash and mineral powder is the largest, with a maximum value of 230×10^{-6}, respectively. Overall, the concrete with a total compounding of 30% and a compounding ratio of 2:1 has the best shrinkage resistance performance.

Keywords: mineral admixture; MEA; fly ash; mineral powder; self-shrinkage; compressive strength

Citation: Zhou, X.; Mao, Z.; Luo, P.; Deng, M. Effect of Mineral Admixtures on the Mechanical and Shrinkage Performance of MgO Concrete. *Materials* **2023**, *16*, 3448. https://doi.org/10.3390/ma16093448

Academic Editor: Weiting Xu

Received: 13 March 2023
Revised: 25 April 2023
Accepted: 26 April 2023
Published: 28 April 2023

Copyright: © 2023 by the authors. Licensee MDPI, Basel, Switzerland. This article is an open access article distributed under the terms and conditions of the Creative Commons Attribution (CC BY) license (https:// creativecommons.org/licenses/by/ 4.0/).

1. Introduction

Concrete is one of the most commonly used construction materials in our life and has superior capacity under compressive loads, but it suffers from some serious defects. These defects are caused by drying shrinkage, thermal shrinkage, self-shrinkage and carbonation shrinkage, which cause the concrete volume to shrink [1,2]. This is the case with high-performance concrete (HPC), which has very good compatibility, mechanical properties and impermeability and is very widely used in engineering. However, the low water-cement ratio and additives make the hydration reaction much faster, which leads to a substantial and rapid temperature drop and a large self-shrinkage [3]. Self-shrinkage of high-performance concrete under certain conditions can lead to excessive concrete stresses, which can lead to cracks [4,5]. Early cracking seriously affects the durability and serviceability and even the safety performance of concrete structures. Therefore, expansion additives are used to compensate for shrinkage, traditionally with sulfate aluminate, aluminate clinker or grass-based [6]. These traditional expansion additives tend to be too dependent on wet curing, which occurs within 14 days of expansion, and early expansion is too fast; they do not compensate for the shrinkage of the concrete after 14 days of construction [7]. Among these expansion agents, MEA has the advantage of stable hydration products and long expansion time and is often used to prepare magnesium-based compensatory shrinkage cement, which is widely used in water conservancy construction [8]. MEA can improve concrete porosity and change concrete pore size distribution [9,10]. In the general mix, MEA content is 3–8%, which can not only compensate for the normal volume shrinkage but also improve the compressive performance, shrinkage performance and durability of concrete [11–14].

For large-volume concrete, with severe shrinkage deformation and high requirements for restriction, it is necessary to incorporate a large amount of MEA, but the delayed expansion of MEA can lead to shrinkage in concrete that is often overcompensated at a later stage, resulting in cracks and a reduction in all properties [15]. To prevent damage to concrete by MEA, some studies tend to reduce the delayed hydration expansion of MEA concrete but ignore the resistance to shrinkage that can enhance the concrete matrix and better coordinate the role of both, which should constrain expansion and compensate for significant shrinkage in the early stages of hydration [16,17]. The faster shrinkage of cement and slower expansion of MEA at ambient conditions can lead to shrinkage stress cracking, so we need to find an auxiliary cementitious material that coordinates the two. Fly ash and mineral powder, as a popular mineral admixture, are very likely to solve this problem depending on their properties in cement [18–20].

In recent years, the increasing cross-sectional size of the component, cement strength level and the amount of cement per unit volume resulted in a significant enhancement in the internal temperature rise during the hardening of concrete. This makes the difference between the standard curing temperature in the laboratory and the temperature in the actual structure, which also leads to a large gap between the strength of the specimen measured in the laboratory and the strength of the concrete in the actual structure. Based on this, this paper investigates the mechanical and shrinkage properties of mineral admixtures (fly ash and mineral powder) on MgO-based concrete in the context of the prevalence of concrete cracking in mass concrete and the actual curing temperature. The changes in expansion and mechanical properties under variable temperature conditions are studied macroscopically, and the effects of mineral admixtures on concrete are analyzed microscopically in terms of pore structure and micromorphology. By simulating the variable temperature environment inside the actual mass concrete, the concrete is mixed and formed, and then cured in the variable temperature environment; the properties of the concrete are measured, its development pattern is grasped, and its mechanism of action is studied, which can be used to guide the actual engineering projects in the future.

2. Materials and Methods

2.1. Raw Materials

The cement selected in this experiment is P·II52.5 Portland cement produced by Onoda of Nanjing, and the strength grade is C50; Fly ash is grade II fly ash of Nanjing Pukou power plant; The Minera powder is S95 grade Minera powder from the Pukou area of Nanjing; The expansion agent is MgO expansion agent produced by Jiangsu Subut New

Materials Co. The fine aggregate is river sand with a fineness modulus of 2.7; the coarse aggregate is crushed stone with a continuous gradation of 5–31.5 mm; the water-reducing agent is polycarboxylic acid high-performance water-reducing agent produced by Jiangsu Subutech New Material Co. The chemical composition of raw materials is shown in Table 1.

Table 1. Chemical compositions of raw materials.

Material	SiO_2 (%)	Al_2O_3 (%)	Fe_2O_3 (%)	CaO (%)	MgO (%)	K_2O (%)	Na_2O (%)	SO_3 (%)	Loss (%)
Cement	18.55	3.95	3.41	65.32	1.01	0.72	0.18	2.78	2.88
Fly ash	44.06	42.06	2.91	3.80	0.40	0.49	0.16	0.75	2.48
Mineral powder	33.39	11.89	0.63	41.51	8.82	0.53	0.67	/	0.28
MEA	3.87	1.03	0.88	1.98	89.37	0.08	/	0.06	2.38

2.2. Sample Preparation

In order to study the effect of mineral admixture on MgO-based concrete under variable temperature conditions, the total amount of cementitious material used is 450 kg/m^3, and the amount of mineral admixture is 30% and 40% of the amount of cementitious material when fly ash is compounded with mineral powder, as this paper studies the effect of the admixture of mineral admixture on the performance of concrete, fly ash is extremely effective in improving the ease of ready-mixed concrete, but the concrete Early and medium-term strength development is slow, mineral powder relative to fly ash on the concrete strength development of each age faster, but poorer than fly ash, the two double admixture of concrete after each age strength and mix compatibility are better, the best ratio of 30% to 40% of the mass of cement [21]. In the compound admixture, the best mass ratio of fly ash to slag powder is 2:1, and the workability of C50 concrete is improved in the process of increasing the amount of external admixture from 0% to 30% [22]. So this paper also selects these two total admixture ratios, where the ratio of fly ash to mineral powder is 3:1, 2:1 or 1:1. MgO expander is mixed internally, and the admixture amount is selected as internal admixture, 8% of the cementitious material, which is higher than the conventional application. The sand rate was 0.39, the water-cement ratio was 0.32, and the water-reducing agent was blended at 2.8% of the cementitious material. The concrete mix ratios are shown in Table 2. The various raw materials were prepared and mixed in a mixer, and then water was added to continue the mixing. After the concrete was mixed, it was poured into a mold of a specific size and removed after one day.

Table 2. Mix proportions of concrete.

Name	Cement (kg/m^3)	Fly Ash (kg/m^3)	Mineral Powder (kg/m^3)	MEA (kg/m^3)	Water (kg/m^3)	Water Reducer
C4-0	315	101.2	33.8	0	144	2.8%
C4	279	101.2	33.8	36	144	2.8%
C5	279	90	45	36	144	2.8%
C6	279	67.5	67.5	36	144	2.8%
C7-0	270	135	45	0	144	2.8%
C7	234	135	45	36	144	2.8%
C8	234	120	120	36	144	2.8%
C9	234	90	90	36	144	2.8%

2.3. Experimental Methods

2.3.1. Heat of Hydration

The temperature difference generated by the heat of hydration released from the hydration of cementitious materials is the main cause of cracks in mass concrete, so a hydration calorimeter (TAM AIR, New Castle, DE, USA) was used to determine the early hydration process of concrete slurry containing MEA. Due to the small water-cement ratio of the cement slurry mixture, it could not be stirred in the calorimeter. Therefore, the calorimeter needs to be placed at a constant temperature of 20 °C for 24 h before testing, and 10.00 g of the test slurry is weighed with an accuracy of 0.01 g and loaded into the test cup of the reaction vessel to continuously follow the exothermic process of hydration of the net cement slurry over a period of time (72 h) to obtain a typical isothermal calorimetric curve, including the exothermic rate and the total exothermic amount with time.

2.3.2. Mechanical Properties

According to Chinese standard GB/T 50081-2019, MgO-based concrete is used in Table 2 ratio; after concrete mixing, molding 150 mm × 150 mm × 150 mm cubic specimens, the molded concrete specimens together with touching tools are wrapped with cling film, input into a large plastic bag, tied with a tie, and put into the variable temperature curing box for curing. Through a whole temperature curve course, after which the specimens are removed and continued to be cured in 20 °C water. The WAW-600C microcomputer controlled electro-hydraulic servo universal testing machine (YiNuo, Jinan, China) was used to measure the 3d, 7d, 28d and 60d compressive strength of concrete. There were 3 concrete specimens in each group, and the compressive strength was taken as its flat value.

Variable temperature curve course as shown in Figure 1, this curve comes from the actual large volume C5 concrete building internal temperature curve measured from the figure, 0–40 h time, the internal temperature of the concrete rose rapidly, reaching a maximum value in the 40thh, the highest temperature of 70 °C, after 40 h temperature began to decline until 18d, the internal temperature of the concrete down to a stable state, continue to measure until 28d, the temperature did not change significantly.

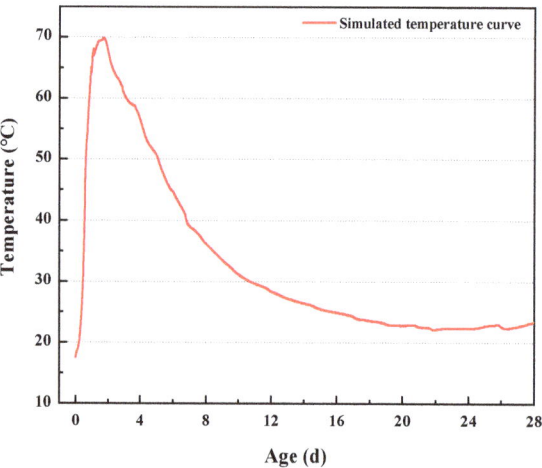

Figure 1. Variable temperature curve process.

2.3.3. Autogenous Deformation

Concrete was mixed and poured in Φ160 mm × 400 mm PVC pipe, pouring in the center of the specimen buried strain gage, and then a slight vibration until no air bubbles were generated. After the completion of the pouring, maintenance 20 °C environment, after the initial setting of the concrete will be closed with paraffin at both ends of the specimen to prevent the influence of water vapor, and then the specimen will go into the variable temperature curing box, continuous monitoring of the deformation of concrete, the concrete specimens are shown in Figure 2. After a whole temperature curve course, remove the specimen and continue curing at 20 °C environment until 28d.

Figure 2. Concrete deformation sample.

2.3.4. Thermal Analysis

The gravel-free concrete mixes were sealed up and placed in a variable temperature curing chamber using ages of 1d, 3d and 7d. The degree of hydration of MgO was evaluated by thermal analysis. The specimens were cut into slices, soaked in anhydrous ethanol for 3d and dried in a vacuum drying oven at 60 °C for 24 h. The specimens were ground into powder and sieved through a 0.08 μm square hole sieve. The DSC-TG method was used to heat the specimens under the N_2 atmosphere with a temperature rise rate of 20 °C/min from 30 °C to 1000 °C.

Since the temperature in the natural environment shows periodic changes, the hydration of MEA will lead to constant changes in the volume of concrete. In order to be able to continuously track the volume deformation of MgO concrete, this experiment uses the MCU-32 automatic measuring instrument produced by Nanjing GeNan Industrial Company in China. To track the strain gauge, strain variables of the test and the acquisition time interval is 1 h. Since the temperature change under natural conditions will cause the volume of concrete to deform, the result calculated by Equation (1) is the strain variation of the concrete specimen after excluding the temperature deformation.

$$\varepsilon_{ml} = k \times \Delta F + (b - \alpha)\Delta T \qquad (1)$$

In Equation (1) ε_{ml} —Strain of concrete after deducting the effect of temperature.

k —Measurement sensitivity of strain gauges (10^{-6}/F).

ΔF —Measurement difference of strain gauges.

b —Thermal expansion coefficient of strain gauge (13.5×10^{-6}/°C).

α —Coefficient of thermal expansion of the concrete specimen under test (5.5×10^{-6}/°C).

2.3.5. Pore Structure

Determination of pore structure and porosity of concrete mixes by Poremaster GT-60 mercury compression meter (Quantachrome, Boynton Beac, FL, USA). The samples of 1d age were cut into small pieces of 2–3 mm in size, put into anhydrous ethanol to terminate hydration, soaked for 3d, and then dried in a vacuum drying oven at 60 °C for 24 h before being taken out for testing. The pore pressure was 415 Mpa, and the pore range was 0.007 μm–360 μm.

2.3.6. SEM Morphology

Samples were taken from the pressed concrete specimens, soaked in anhydrous ethanol to terminate hydration, soaked for 3d, dried at a temperature of 60 °C for 24 h and then removed, the specimens were coated with gold, and the microscopic morphology of concrete was observed using a JSM6480 scanning electron microscope(Thermo Fisher Scientific, Waltham, MA, USA), and the relationship between strength and concrete structural densities was desired.

3. Results and Discussion
3.1. Hydration Heat

Figure 3 shows the heat flow in 3 d for different mineral admixtures. The results show that the induction period occurs mainly within 2 h. The accelerated period has a faster exothermic rate and lasts for about 11 h. The accelerated and delayed periods take a total of 13 h, which is basically consistent with the initial setting time of concrete. The longer induction time is due to the fact that less free water in the system inhibits the diffusion of Ca^{2+} and OH^-. On the other hand, the high concentration of SP also inhibited the nucleation growth of unhydrated cement particles and the formation of $Ca(OH)_2$, which led to a slower hydration process in the HPC system [23]. It is presumed that the early expansion of concrete may be caused by the temperature increase; after that, as the hydration of the system enters the deceleration period, which lasts about 20 h, at the early stage of deceleration, the exothermic heat of hydration is greater than the exothermic heat, so the internal temperature of concrete is still rising until the maximum temperature, and when the exothermic heat is smaller than the exothermic heat, the temperature starts to fall until the room temperature. It can be presumed that concrete accelerates expansion in the early stage of hydration and starts shrinking in the late stage of hydration. Eventually, the whole system of cementitious material hydration is basically stable, and the heat dissipation is small.

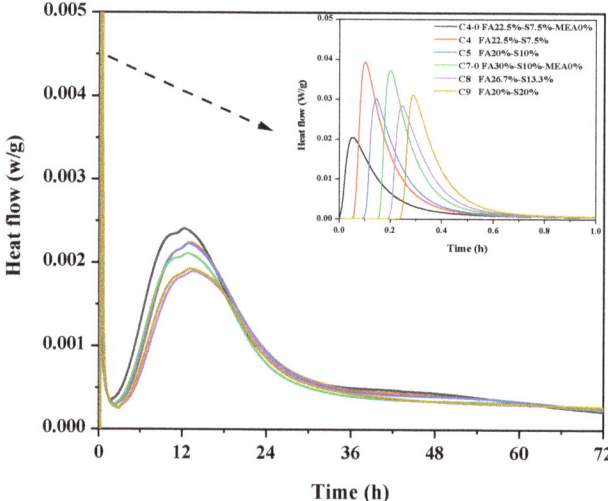

Figure 3. Effect of mineral admixtures on heat flow rate.

As shown in Figure 3, C4 leads to a slowing down of the heat flow of hydration when 8% MEA is incorporated compared to C4-0. Additionally, increasing the total amount of mineral admixture to 40% did not show a longer induction period and acceleration cycle but only a significant reduction in the hydration heat flow. As seen in the figure, changing the admixture of fly ash and mineral powder thus leads to a change in the release of hydration heat flow; when the admixture of fly ash is 20% and the admixture of mineral powder is 20%, the heat release peak is significantly wider and shows longer induction period and acceleration period. For MgO-based shrinkage-compensated cement, the incorporation of mineral admixture enhances the aluminate reaction and diminishes the effect of MEA in accelerating the thermodynamic reaction [24]. Therefore, changing the ratio of mineral admixtures can reduce the possibility of MgO-based large-volume concrete cracking due to temperature stresses.

3.2. Autogenous Shrinkage

Figure 4 depicts autogenous shrinkage of fly ash and S95 ore pulverized concrete after deducting the influence of temperature. Figure 4a depicts the self-deformation of the concrete compounded with 30% fly ash and S95 mineral powder in 28d net of temperature effects. The red line in the figure shows the temperature change curve for the whole test, which shows that the test reaches the highest temperature at 1d, about 75 °C, and then proceeds to a slight decrease to room temperature of 20 °C at 18d. All samples were expanding rapidly during the early temperature rise period, but the C4-0 sample showed a lack of subsequent expansion, which was smaller than that of the MEA-doped sample, and eventually began to shrink after experiencing the highest temperature and showed contraction after 28d. The early expansion of the C6 sample was the largest, indicating that fly ash had an inhibitory effect on early expansion under variable temperature conditions. The samples doped with MEA eventually exhibited expansion after 28d, which indicates that MEA compensates well for early shrinkage.

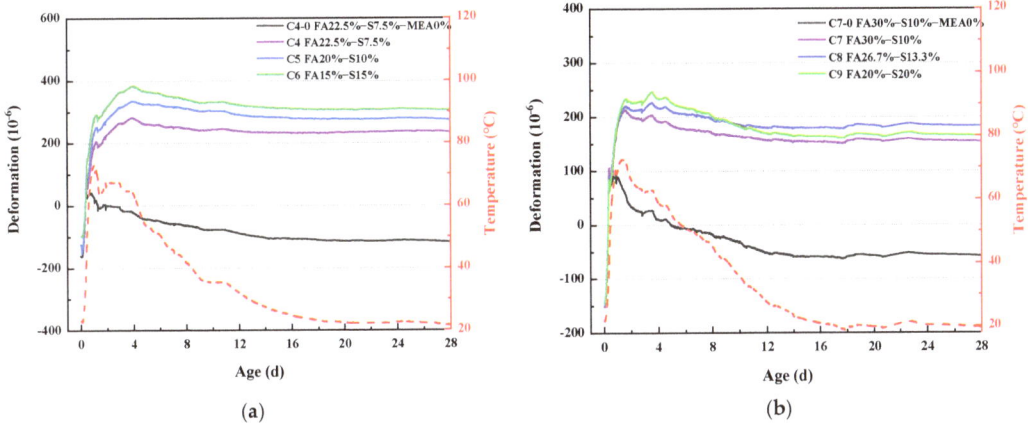

Figure 4. Autogenous Shrinkage of fly ash and S95 ore pulverized concrete after deducting the influence of temperature. (**a**) Total dosage is 30%; (**b**) Total dosage is 40%.

Figure 4b depicts the self-deformation of the concrete compounded with 40% fly ash and S95 mineral powder in 28 days net of temperature effects. The C9 samples started with greater volume expansion than C8 and C7, and subsequently, the generated $Ca(OH)_2$ and $Mg(OH)_2$ stimulated the volcanic ash activity of the fly ash, producing large amounts of C-S-H, C-A-H, and M-S-H [25,26]. The maximum volume deformation of concrete with 30% secondary fly ash and 10% mineral powder without MEA is 90×10^{-6}, and the stable value is -50×10^{-6}, showing a shrinkage state. In the proportion of concrete mixed with MEA, the maximum volume deformation of concrete mixed with 30% second-grade fly ash and 10% mineral powder is 220×10^{-6}, and the stable value is 150×10^{-6}. The maximum volume deformation of concrete with 26.7% second-grade fly ash and 13.3% mineral powder is 230×10^{-6}, and the stable value is 180×10^{-6}. The maximum volume deformation of concrete with 20% secondary grade fly ash and 20% mineral powder is 250×10^{-6}, and the stable value is 170 × 10⁻⁶. It can be seen from the above data that the concrete mixed with 26.7 secondary fly ash and 13.3% mineral powder has the smallest shrinkage with the decrease in temperature, and the concrete has the best shrinkage resistance.

3.3. Compressive Strength

Figure 5a depicts the compressive strengths at 3d, 7d, 28d and 60d for concrete compounded with 30% fly ash and S95 mineral powder. The compressive strength gradually decreases with increasing the ratio of fly ash to mineral powder 3:1, 2:1 and 1:1, and the incorporation of MEA also leads to a decrease in compressive strength. For the mineral admixture concrete system, the compressive strength of C4, C5 and C6 samples at 3d age decreased by 2.76 Mpa, 3.3 Mpa and 5.6 Mpa, respectively, compared to C4-0 samples, while the compressive strength of C4, C5 and C6 samples at 60d age decreased by 0.6 Mpa, 2.3 Mpa and 3 Mpa, respectively, compared to C4-0 samples. The effect was greater at the early stage and had almost No effect. It was found that the adverse effect on the compressive strength of concrete may be due to the fact that MEA reduces the volume fraction of cement, while hydration products, such as magnesium hydroxide, contribute less to the strength [27]. In addition, it is also possible that the hydration products have relatively small crystals and are less strong than the cement hydration products, resulting in lower mechanical properties [28].

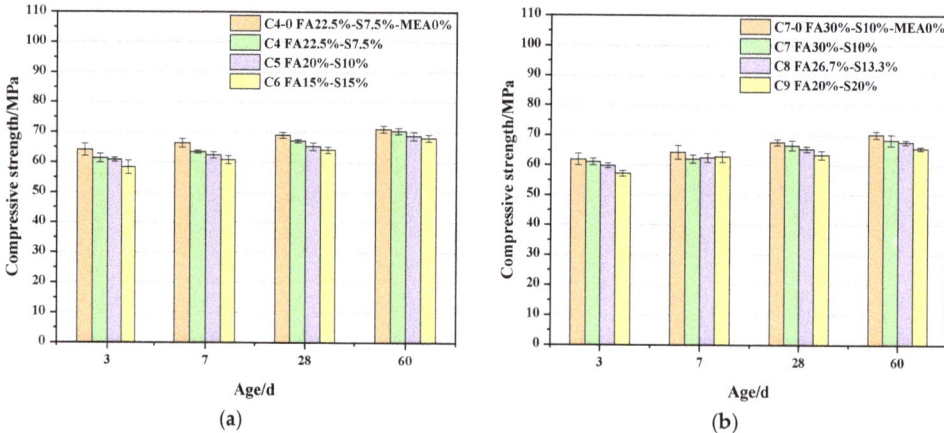

Figure 5. Compressive strength of fly ash and S95 ore pulverized concrete. (**a**) Total dosage is 30%; (**b**) Total dosage is 40%.

Figure 5b shows the histogram of compressive strength with age for concrete with 40% total mineral admixture under variable temperature conditions. We found that the compressive strength of concrete mixed with MEA was generally higher than that of the unmixed 40% mineral admixture, but unlike the 30% mineral admixture, the compressive strength of concrete at age 7d surprisingly increased with the decrease of fly ash admixture and the increase of mineral powder admixture. The compressive strength at the 3d age of concrete mixed with MEA decreased by 1.3%, 3.4% and 7.6%, respectively, compared with that of unmixed. The compressive strength of 3d concrete decreased with decreasing amount of secondary fly ash and an increasing amount of mineral powder. 3d compressive strength of concrete mixed with 30% secondary fly ash and 10% mineral powder was the highest in concrete mixed with MEA. 28d age, for the blank group, they decreased by 1.5%, 3.3% and 6.2%, respectively. At the age of 60d, the compressive strength of the concrete mixed with MEA was lower than that of the blank group by 2.6%, 3.6% and 6.6%, respectively. It indicates that the massive admixture of fly ash has a positive effect on the development of concrete strength under variable temperature conditions.

3.4. XRD

Figure 6 describes the XRD patterns of C5, C4, C8 and C9 samples at 1d, 3d and 7d. The unhydrated magnesite in C4 samples exceeded C5 on day 7. With the increase of fly ash content, the unhydrated magnesite content in cement slurry increased, indicating that fly ash inhibited the hydration of MEA at the early stage of the variable temperature environment. The unhydrated cubic magnesite in the C8 sample exceeded C5 on the 7th day, which was also the concrete with a 2:1 dosage ratio. The content of unhydrated cubic magnesite in the cement slurry with a total content of 30% was lower than that in the cement slurry with a total content of 40%, indicating that mineral admixtures could inhibit the hydration of MEA. In the MgO-SiO$_2$-H$_2$O system, MgO can not only be chemically synthesized into Mg(OH)$_2$ but also react with SiO$_2$ to form M-S-H. Among them, the energy required to complete the second reaction is less than the first one [29]. With the addition of fly ash, the CaO content decreases and the SiO$_2$ content increases. This way, the residual SiO$_2$ combines with MgO and Mg(OH)$_2$ as much as possible to form M-S-H [30–34].

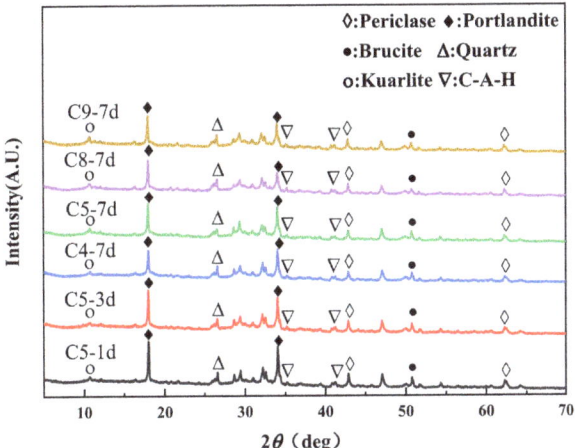

Figure 6. XRD spectra of C5, C4, C8 and C9.

3.5. Hydration Degree of MgO

The TG curves of C4, C5, C6, C7, C8 and C9 samples at d 1d, 3d and 7d are shown in Figure 7. In the temperature range of 30–330 °C, some hydration products, for example, C-A-H, C-S-H and Afm, decompose when heated to a certain temperature [35]. What we need to understand is that Mg(OH)$_2$ and Ca(OH)$_2$ undergo thermal decomposition at a temperature range of 330–420 °C and 420–500 °C, respectively [36,37].

The equations of the calculation equation are displayed in Equations (2) and (3). The fraction of the mass of the sample at 1d age under variable temperature conditions is displayed in Table 3. Therefore, the content of Mg(OH)$_2$ can be calculated according to Equation (4) and Table 3, and the results are shown in Figure 8.

$$\mathrm{Mg(OH)_2} \rightarrow \mathrm{MgO} + \mathrm{H_2O} \tag{2}$$

$$\mathrm{Ca(OH)_2} \rightarrow \mathrm{CaO} + \mathrm{H_2O} \tag{3}$$

$$W_{\mathrm{Mg(OH)_2}} = \Delta W_{330-420\,°C} \frac{M_{\mathrm{Mg(OH)_2}}}{M_{\mathrm{H_2O}}} \tag{4}$$

In Equations (2)–(4) $W_{Mg(OH)_2}$ is the mass fraction of $Mg(OH)_2$ (%); $M_{Mg(OH)_2}$ and M_{H_2O} are the molar masses of $Mg(OH)_2$, H_2O (g/mol). $\Delta W_{330-420°C}$ expresses the mass loss fraction in the temperature range of 330–420 °C.

Figure 8 shows the content of $Mg(OH)_2$ in the pure pulp of Grade II fly ash and mineral powder cement mixed with 30% and 40% at variable temperatures. It can be seen from the figure that, The $Mg(OH)_2$ content of concrete slurry with 22.5% secondary fly ash, 7.5% ore powder, 20% secondary fly ash 10% ore powder, 15% secondary fly ash, 15% ore powder, 30% secondary fly ash 10% ore powder, 26.7% secondary fly ash 13.3% ore powder and 20% secondary fly ash 20% ore powder at 7 days age, respectively is 9.24%, 9.51%, 9.86%, 8.60%, 8.60% and 9.73%. 15% secondary fly ash has the largest $Mg(OH)_2$ content and corresponding deformation, with an expansion of 350 με at 7d, while 30% secondary fly ash 10% mineral powder has the lowest $Mg(OH)_2$ content and the lowest deformation, with an expansion of 170 με at 7d. The curve in the figure also decreases with the increase of secondary fly ash content and the decrease of mineral powder content. For example, the C6 curve in the figure is above C5. It can be seen that the incorporation of fly ash weakens the hydration process of magnesium oxide to some extent. Although the $Mg(OH)_2$ content of 15% secondary fly ash and 20% secondary fly ash is very similar at 3d age, the expansion of the former is 120 με larger than that of the latter. From the point of view of total shrinkage compensation, the concrete mixed with 20% secondary fly ash and 10% mineral powder is the most suitable.

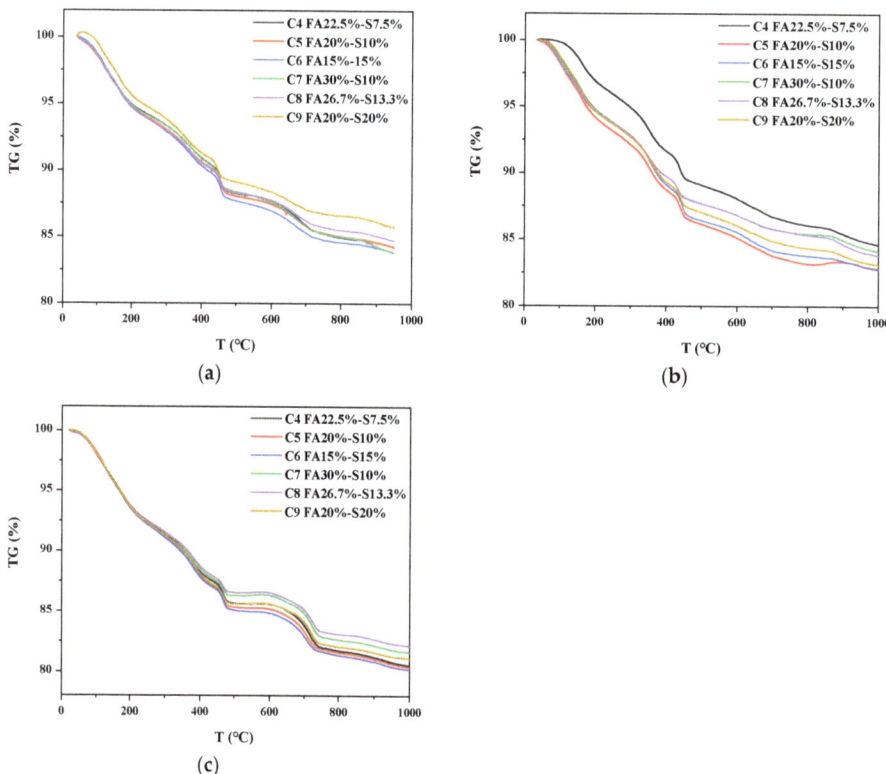

Figure 7. Hydration product contents at different ages. (**a**) 1 day of age; (**b**) 3 days of age; (**c**) 7 days of age.

Table 3. Mass fraction of 1-day-old samples at different temperatures.

Name	C4	C5	C6	C7	C8	C9
330 °C	92.67%	92.36%	92.20%	92.67%	92.32%	93.07%
420 °C	90.53%	90.23%	89.91%	90.46%	93.18%	90.95%

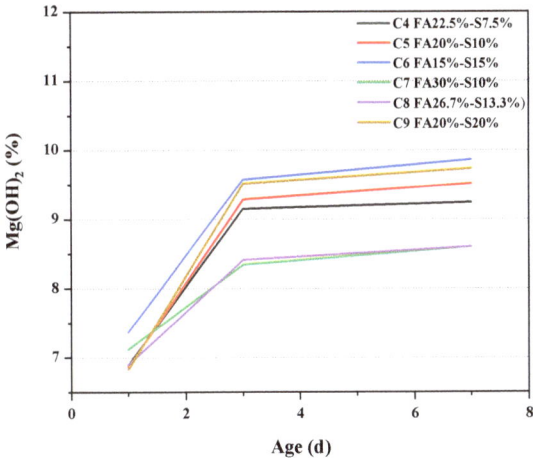

Figure 8. The content of $Mg(OH)_2$ at different ages.

3.6. Microstructure

3.6.1. Pore Structure

Figure 9a depicts the pore structure of the different mineral admixtures at the age of 7d. It can be seen that the pore sizes of the samples are mainly distributed in the range of 0.007 to 0.03 μm. Compared with the C5 sample, the pore size peak of the C9 sample is significantly shifted to the right, indicating that the multiple admixtures of mineral powder lead to larger pores of the whole system. During the admixture of fly ash, the pores become less, which will make the concrete denser and strengthen the mechanical properties of concrete. It indicates that fly ash can stabilize the MEA and improve structural compactness. As in Figure 9b, the magnitude of porosity in the samples is C4, C5 < C9, indicating that the increase in the total admixture of mineral admixture leads to an increase in the porosity of cement paste in concrete but a significant decrease in the average pore size. Increasing the admixture of mineral powder also leads to an increase in porosity, which is not conducive to the stability of the concrete structure. Overall, the distribution curve of pore size is generally shifted towards smaller pore sizes. Therefore, a larger proportion of fly ash can mitigate the self-shrinkage of concrete more effectively.

3.6.2. SEM

In order to have a deeper understanding of the differences in the mechanical strength development of C4 and C6 samples from 3d to 60d, Figure 10 shows the SEM observation. There is no flocculation on the surface of C4 samples at 60 days, and there is no fracture layer at the links, which indicates that the multiple admixtures of fly ash have a positive effect on the later tissue growth of C5 samples. While SEM of the C6 sample at 60 days, flocs were formed on the surface, and these flocs were flake structures, and this structure was not conducive to the development of concrete strength. The loose structure of these $Mg(OH)_2$ crystals is related to the hydration of MgO, which leads to premature swelling. Strongly demonstrates the positive effect of fly ash on improving the structural compactness of concrete. There are two possible reasons for this positive effect. One is that fly ash attenuates the delayed over-swelling by reducing Mg(OH)$_2$, and the other is that it forms a better cement matrix.

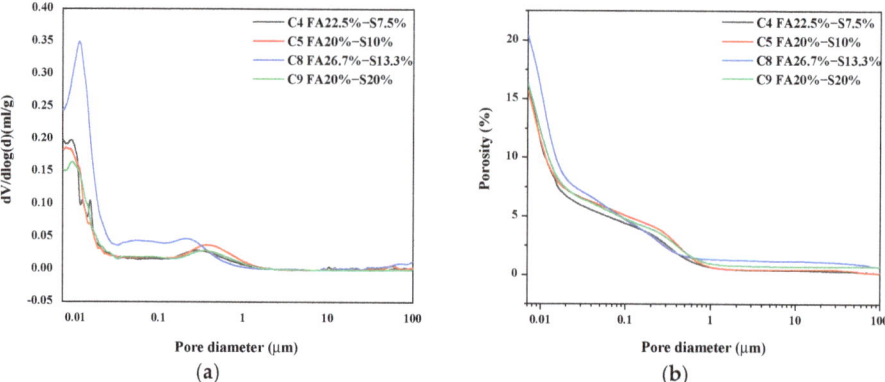

Figure 9. Pore size structure of slurries with different mineral admixtures. (**a**) Total dosage is 30%; (**b**) Total dosage is 40%.

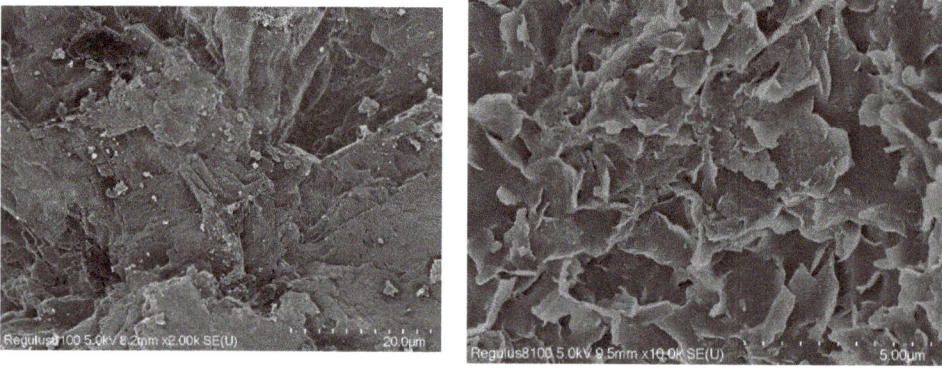

Figure 10. SEM images of test specimens.

4. Conclusions

In this topic, the effects of mineral admixtures on MgO concrete under variable temperature conditions were systematically studied from some aspects, such as shrinkage properties, mechanical properties and microstructure. Based on the experimental results, the following conclusions can be obtained.

1. Compared with the sample without MEA, the addition of MEA can significantly reduce the heat flow of cement, increase the mineral admixture can, reduce the heat of hydration and inhibit the delayed overexpansion of concrete, improve the later mechanical properties of concrete, and reduce the risk of cracking of concrete.
2. The results of the self-shrinkage show that concrete expands sharply and then shrinks under variable temperature conditions. However, concrete without MEA eventually exhibits shrinkage, while concrete with MEA exhibits expansion. The incorporation of mineral admixtures prevents cracking of the concrete due to self-shrinkage and prevents delayed over-expansion of the concrete in the later stages.
3. Under variable temperature conditions, the addition of mineral admixtures will lead to a decrease in the mechanical properties of concrete, but the amount of reduction is very small and meets the actual needs. Under variable temperature conditions, the

mechanical properties decrease with the decrease of fly ash content, indicating that the incorporation of fly ash is conducive to the mechanical properties of concrete.
4. The XRD and TG results show that under variable temperature conditions, fly ash leads to weaker shrinkage resistance of concrete in the early stages by reducing the hydration of MEA inside the concrete, and in the later stages, mineral admixtures form stronger concrete cementing substances through secondary reactions. Overall, fly ash can harmonize the relationship between shrinkage resistance and shrinkage of concrete.
5. MIP analysis shows that the porosity of the C5 sample with 10% ore powder of 20% fly ash is 16%, while the porosity of 20% ore powder of 20% fly ash is 17%. Increasing the ore powder content can increase the porosity, indicating that a large amount of ore powder is not conducive to the thinning of pore size. The addition of fly ash makes the pore size refined continuously, the porosity reduced, and the hydration products closer to each other. It will make the structure of concrete more dense and strengthen the mechanical properties of the content.

5. Recommendation

By studying the laws of mineral admixtures on the mechanical and shrinkage properties of concrete under variable temperature conditions. When working on mass concrete, we need to fully consider its internal temperature rise due to the hydration of cement, where temperature gradient changes can lead to cracking of the concrete, thus, affecting the durability performance of concrete and even safety hazards. In order to reduce the temperature rise of concrete, we can take the internal buried cooling pipes and also add mineral admixtures. The above study found that adding 30% of mineral admixture is the most suitable, and 30% of the total amount of compounding, compounding ratio 2:1, and the best case, its final deformation for the expansion state, to solve the cracking of concrete because of shrinkage, also will not occur excessive expansion of concrete, its mechanical properties are also to meet the project requirements.

Author Contributions: Conceptualization, M.D.; data curation, X.Z., Z.M. and P.L.; writing—original draft preparation, X.Z.; writing—review and editing, M.D., P.L. and Z.M. All authors have read and agreed to the published version of the manuscript.

Funding: This work was supported by the Open Fund project of the National Laboratory for High-Performance Civil Engineering Materials (2021CEM012) and the Priority Academic Program Development of Jiangsu Higher Education Institutions (PAPD).

Institutional Review Board Statement: Not applicable.

Informed Consent Statement: Not applicable.

Data Availability Statement: The data presented in this study are available on request from the corresponding author.

Acknowledgments: The authors gratefully acknowledge the assistance from Zhongyang Mao from NJTECH and the staff from the State Key Laboratory of Materials-Oriented Chemical Engineering.

Conflicts of Interest: The authors declare no conflict of interest.

References

1. Barr, B.; Hoseinian, S.; Beygi, M. Shrinkage of concrete stored in natural environments. *Cem. Concr. Compos.* **2003**, *25*, 19–29. [CrossRef]
2. West, R.P.; Holmes, N. Predicting moisture movement during the drying of concrete floors using finite elements. *Constr. Build. Mater.* **2005**, *19*, 674–681. [CrossRef]
3. Zhao, H.; Jiang, K.; Hong, B.; Yang, R.; Xu, W.; Tian, Q.; Liu, J. Experimental and Numerical Analysis on Coupled Hy-gro-Thermo-Chemo-Mechanical Effect in Early-Age Concrete. *J. Mater. Civ. Eng.* **2021**, *33*, 04021064. [CrossRef]
4. Li, W.; Huang, Z.; Hu, G.; Duan, W.H.; Shah, S.P. Early-age shrinkage development of ultra-high-performance concrete under heat curing treatment. *Constr. Build. Mater.* **2017**, *131*, 767–774. [CrossRef]

5. Li, M.; Xu, W.; Wang, Y.; Tian, Q.; Liu, J. Shrinkage crack inhibiting of cast in situ tunnel concrete by double regulation on temperature and deformation of concrete at early age. *Constr. Build. Mater.* **2019**, *240*, 117834. [CrossRef]
6. Mo, L.; Deng, M.; Tang, M.; Al-Tabbaa, A. MgO expansive cement and concrete in China: Past, present and future. *Cem. Concr. Res.* **2014**, *57*, 1–12. [CrossRef]
7. Gao, P.; Lu, X.; Geng, F.; Li, X.; Hou, J.; Lin, H.; Shi, N. Expansive admixtures (mainly ettringite). *Cem. Concr. Compos.* **1998**, *20*, 163–170. [CrossRef]
8. Gao, P.W.; Lu, X.L.; Geng, F.; Li, X.; Hou, J.; Lin, H.; Shi, N. Production of MgO-type expansive agent in dam concrete by use of industrial by-products. *Build. Environ.* **2008**, *43*, 453–457. [CrossRef]
9. Choi, S.-W.; Jang, B.-S.; Kim, J.-H.; Lee, K.-M. Durability characteristics of fly ash concrete containing lightly-burnt MgO. *Constr. Build. Mater.* **2014**, *58*, 77–84. [CrossRef]
10. Gao, P.-W.; Wu, S.-X.; Lin, P.-H.; Wu, Z.-R.; Tang, M.-S. The characteristics of air void and frost resistance of RCC with fly ash and expansive agent. *Constr. Build. Mater.* **2006**, *20*, 586–590. [CrossRef]
11. Xue, C.; Li, W.; Luo, Z.; Wang, K.; Castel, A. Effect of chloride ingress on self-healing recovery of smart cementitious composite incor-porating crystalline admixture and MgO expansive agent. *Cem. Concr. Res.* **2021**, *139*, 106252. [CrossRef]
12. Li, W.; Huang, Z.; Zu, T.; Shi, C.; Duan, W.H.; Shah, S.P. Influence of Nanolimestone on the Hydration, Mechanical Strength, and Autogenous Shrinkage of Ultrahigh-Performance Concrete. *J. Mater. Civil. Eng.* **2016**, *28*, 04015068. [CrossRef]
13. Zhang, J. Recent advance of MgO expansive agent in cement and concrete. *J. Build. Eng.* **2022**, *45*, 103633. [CrossRef]
14. Li, C. Review of quick damming technology of MgO concrete. *Adv. Sci. Technol. Water Resour.* **2013**, *33*, 82–88.
15. Amaral, L.; Oliveira, I.; Salomão, R.; Frollini, E.; Pandolfelli, V. Temperature and common-ion effect on magnesium oxide (MgO) hydration. *Ceram. Int.* **2010**, *36*, 1047–1054. [CrossRef]
16. Jiang, D.; Li, X.; Lv, Y.; Li, C.; Zhang, T.; He, C.; Leng, D.; Wu, K. Early-age hydration process and autogenous shrinkage evolution of high performance cement pastes. *J. Build. Eng.* **2022**, *45*, 103436. [CrossRef]
17. Qian, G.; Xu, G.; Li, H.; Li, A. The effect of autoclave temperature on the expansion and hydrothermal products of high-MgO blended cements. *Cem. Concr. Res.* **1998**, *28*, 1–6. [CrossRef]
18. Deschner, F.; Winnefeld, F.; Lothenbach, B.; Seufert, S.; Schwesig, P.; Dittrich, S.; Goetz-Neunhoeffer, F.; Neubauer, J. Hydration of Portland cement with high replacement by siliceous fly ash. *Cem. Concr. Res.* **2012**, *42*, 1389–1400. [CrossRef]
19. Bentz, D.P. Powder Additions to Mitigate Retardation in High-Volume Fly Ash Mixtures. *ACI Mater. J.* **2010**, *107*, 508–514.
20. Hanif, A.; Lu, Z.; Li, Z. Utilization of fly ash cenosphere as lightweight filler in cement-based composites—A review. *Constr. Build. Mater.* **2017**, *144*, 373–384. [CrossRef]
21. Li, Y.; Jia, L.; Ma, Q. Influence of fly ash, slag powder on the workability and strength of low-grade concrete. *Concrete* **2014**, *75*, 43–52.
22. Xia, L.; Ni, T.; Liu, Z. Effect of Mineral Admixture on Workability and Mechanical Properties of Box Girder C50 Concrete. *Bull. Chin. Ceram. Soc.* **2017**, *36*, 1193–1197.
23. Jansen, D.; Neubauer, J.; Goetz-Neunhoeffer, F.; Haerzschel, R.; Hergeth, W.D. Change in reaction kinetics of a Portland cement caused by a super-plasticizer—Calculation of heat flow curves from XRD data. *Cem. Concr. Res.* **2012**, *42*, 327–332. [CrossRef]
24. Zhang, J.; Lv, T.; Han, Q.; Zhu, Y.; Hou, D.; Dong, B. Effects of fly ash on MgO-based shrinkage-compensating cement: Microstructure and properties. *Constr. Build. Mater.* **2022**, *339*, 127648. [CrossRef]
25. Mo, L.; Liu, M.; Al-Tabbaa, A.; Deng, M. Deformation and mechanical properties of the expansive cements produced by in-ter-grinding cement clinker and MgOs with various reactivities. *Constr. Build. Mater.* **2015**, *80*, 1–8. [CrossRef]
26. Chen, C.L.; Fang, K.H.; Jiang, J. Influence of fly ash content on autogenic volume change of concrete mixed with MgO. *Concrete* **2012**, *29*, 67–69.
27. Li, Y.; Deng, M.; Mo, L.; Tang, M. Strength and Expansive Stresses of Concrete with MgO Type Expansive Agent under Restrain Conditions. *J. Build. Mater.* **2012**, *15*, 446–450.
28. Li, S.; Mo, L.; Deng, M.; Cheng, S. Mitigation on the autogenous shrinkage of ultra-high performance concrete via using MgO ex-pansive agent. *Constr. Build. Mater.* **2021**, *312*, 125422. [CrossRef]
29. Li, Z.; Zhang, T.; Hu, J.; Tang, Y.; Niu, Y.; Wei, J.; Yu, Q. Characterization of reaction products and reaction process of MgO-SiO$_2$-H$_2$O system at room temperature. *Constr. Build. Mater.* **2014**, *61*, 252–259. [CrossRef]
30. Dung, N.; Unluer, C. Carbonated MgO concrete with improved performance: The influence of temperature and hydration agent on hydration, carbonation and strength gain. *Cem. Concr. Compos.* **2017**, *82*, 152–164. [CrossRef]
31. Filippou, D.; Katiforis, N.; Papassiopi, N.; Adam, K. On the kinetics of magnesia hydration in magnesium acetate solutions. *J. Chem. Technol. Biotechnol.* **1999**, *74*, 322–328. [CrossRef]
32. Dung, N.T.; Unluer, C. Sequestration of CO$_2$ in reactive MgO cement-based mixes with enhanced hydration mechanisms. *Constr. Build. Mater.* **2017**, *143*, 71–82. [CrossRef]
33. Matabola, K.P.; van der Merwe, E.M.; Strydom, C.A.; Labuschagne, F.J.W. The influence of hydrating agents on the hydration of industrial magnesium oxide. *J. Chem. Technol. Biotechnol.* **2010**, *85*, 1569–1574. [CrossRef]
34. Kuenzel, C.; Zhang, F.; Ferrandiz-Mas, V.; Cheeseman, C.R.; Gartner, E.M. The mechanism of hydration of MgO-hydromagnesite blends. *Cem. Concr. Res.* **2018**, *103*, 123–129. [CrossRef]
35. Alarcon-Ruiz, L.; Platret, G.; Massieu, E.; Ehrlacher, A. The use of thermal analysis in assessing the effect of temperature on a cement paste. *Cem. Concr. Res.* **2005**, *35*, 609–613. [CrossRef]

36. Bahafid, S.; Ghabezloo, S.; Duc, M.; Faure, P.; Sulem, J. Effect of the hydration temperature on the microstructure of Class G cement: C-S-H composition and density. *Cem. Concr. Res.* **2017**, *95*, 270–281. [CrossRef]
37. Zhao, Z.; Qu, X.; Li, J. Microstructure and properties of fly ash/cement-based pastes activated with MgO and CaO under hydrothermal conditions. *Cem. Concr. Compos.* **2020**, *114*, 103739. [CrossRef]

Disclaimer/Publisher's Note: The statements, opinions and data contained in all publications are solely those of the individual author(s) and contributor(s) and not of MDPI and/or the editor(s). MDPI and/or the editor(s) disclaim responsibility for any injury to people or property resulting from any ideas, methods, instructions or products referred to in the content.

Article

Effect of Hydration Temperature Rise Inhibitor on the Temperature Rise of Concrete and Its Mechanism

Tian Liang [1], Penghui Luo [1,2], Zhongyang Mao [1,2], Xiaojun Huang [1,2], Min Deng [1,2] and Mingshu Tang [1,2,*]

[1] College of Materials Science and Engineering, Nanjing Tech University, Nanjing 211800, China; 202061203173@njtech.edu.cn (T.L.); dengmin@njtech.edu.cn (M.D.)

[2] State Key Laboratory of Materials-Oriented Chemical Engineering, Nanjing 211800, China

* Correspondence: tangmingshu@njtech.edu.cn; Tel.: +86-136-0518-4865

Abstract: The rapid drop in internal temperature of mass concrete can readily lead to temperature cracks. Hydration heat inhibitors reduce the risk of concrete cracking by reducing the temperature during the hydration heating phase of cement-based material but may reduce the early strength of the cement-based material. Therefore, in this paper, the influence of commercially available hydration temperature rise inhibitors on concrete temperature rise is studied from the aspects of macroscopic performance and microstructure characteristics, and their mechanism of action is analyzed. A fixed mix ratio of 64% cement, 20% fly ash, 8% mineral powder and 8% magnesium oxide was used. The variable was different admixtures of hydration temperature rise inhibitors at 0%, 0.5%, 1.0% and 1.5% of the total cement-based materials. The results showed that the hydration temperature rise inhibitors significantly reduced the early compressive strength of concrete at 3 d, and the greater the amount of hydration temperature rise inhibitors, the more obvious the decrease in concrete strength. With the increase in age, the influence of hydration temperature rise inhibitor on the compressive strength of concrete gradually decreased, and the decrease in compressive strength at 7 d was less than that at 3 d. At 28 d, the compressive strength of the hydration temperature rise inhibitor was about 90% in the blank group. XRD and TG confirmed that hydration temperature rise inhibitors delay early hydration of cement. SEM showed that hydration temperature rise inhibitors delayed the hydration of $Mg(OH)_2$.

Keywords: hydration heat inhibitor; heat of hydration; concrete; degree of hydration

Citation: Liang, T.; Luo, P.; Mao, Z.; Huang, X.; Deng, M.; Tang, M. Effect of Hydration Temperature Rise Inhibitor on the Temperature Rise of Concrete and Its Mechanism. *Materials* 2023, 16, 2992. https://doi.org/10.3390/ma16082992

Academic Editor: Alessandro P. Fantilli

Received: 17 March 2023
Revised: 2 April 2023
Accepted: 5 April 2023
Published: 10 April 2023

Copyright: © 2023 by the authors. Licensee MDPI, Basel, Switzerland. This article is an open access article distributed under the terms and conditions of the Creative Commons Attribution (CC BY) license (https:// creativecommons.org/licenses/by/ 4.0/).

1. Introduction

After the mass concrete is poured, with the cement hydration and heat release, the internal temperature of the mass concrete will change sharply. With the increase in curing time, the strength of concrete increases [1], and the structure is constrained (foundation constraint, new and old concrete contact surface constraint, etc.). Uneven temperature distribution and different constraints within concrete will lead to large temperature stress [2,3]. Once the temperature stress exceeds the permissible tensile strength of concrete, temperature cracks will occur [4,5]. This not only affects the function of the building structure and reduces its stiffness but also affects the durability of the concrete. It has a great negative effect on the mechanical properties and application properties of concrete. Studies show that, with the increase in temperature, the hydration heat release rate of C_3S accelerates significantly [6]. Therefore, the commonly used measures to control the heat of hydration include using low-heat cement, increasing aggregate content during mixing, controlling the concrete placement temperature and cooling with water [7]. In addition, concrete admixture is also an effective means to improve the thermodynamic properties of concrete, and this method has been widely used in engineering [8,9]. From then on, the choice of concrete admixture is the key to solve the problem of concrete cracking. Hydration heat inhibitors [10–12] is a new type of concrete admixture, prepared by acid hydrolysis of corn starch. It is used to solve the problem of temperature cracking caused by excessive thermal stress in mass concrete [13,14]. Studies have shown that adding hydration heat inhibitors to concrete can effectively reduce the heat release rate of cement in the hydration acceleration stage without affecting the total heat release of hydration. That is, temperature rise inhibitors can regulate the hydration process and reduce the heat release rate of hydration [15,16]. The main components of existing hydration heat inhibitors are hydroxyl carboxylic acid esters, starch or dextrin and its derivatives, active heat-absorbing salts, etc. In addition, the hydration heat inhibitor can be modified to form a new type of hydration heat inhibitor. Relevant studies [15] have shown that the addition of hydration heat inhibitors has a significant impact on the nucleation of C-S-H, which is mainly manifested as reducing the main peak of hydration reaction. Hydration heat inhibitors cement hydration mainly by regulating the induction and acceleration time of C_3S hydration, and the hydration heat inhibitors' effect on the early stage of cement hydration is greater than that in the later stage because, once the nucleus is formed, the growth of C-S-H will be very stable and will not be affected by hydration heat inhibitors. Many domestic and foreign scholars have studied the effect of hydration heat inhibitors on cement hydration [17] and concrete hydration [11]. These studies basically focus on the effects of hydration heat inhibitors on pre-hydration or post-hydration, and they often ignore the coordination between the two. Specifically, hydration heat inhibitors can reduce the early strength of concrete, delay the concrete setting time, etc. This limits its application in engineering cast-in-place structural concrete. Therefore, it is necessary to develop a hydration heat inhibitor that has less influence on the setting time and early strength of concrete.

In this paper, the effects of adding different hydration heat inhibitor dosages on cement concrete were studied from macroscopic properties and microscopic structure characteristics. Macroscopic performance shows that the addition of hydration heat inhibitor has the effect of delaying cement hydration and can relatively reduce the internal temperature rise of concrete and delay the temperature rise. Based on the characteristics of microstructure, the mechanism of hydration heat inhibitor on concrete was analyzed from the aspects of phase composition, pore structure and morphology.

2. Materials and Methods
2.1. Materials

The cement selected for the experiment was Nanjing Onoda PII 52.5 silicate cement; the fly ash was secondary ash; the mineral powder was S95 grade mineral powder; the expansion agent was MgO expansion agent produced by Jiangsu Subote New Materials Joint Stock Company (Zhenjiang, China); the fine aggregate was river sand with a fineness modulus of 2.7; the coarse aggregate was 5–31.5 mm continuously graded crushed stone; the water reducing agent was poly(carboxylic acid) high-performance water reducing agent produced by Jiangsu Subote New Materials Joint Stock Company. The water reducing agent is poly-carboxylic acid high-performance water reducing agent produced by Jiangsu Subote New Materials Co. (Nanjing, China) The water reducing agent is the high-performance water reducing agent of polycarboxylic acid produced by Jiangsu Subote New Materials Co. Hydration heat inhibitor is the product of Jiangsu Subote New Materials Co., hereinafter referred to as SBT.

The chemical composition of the raw material is shown in Table 1. The functional groups test of the SBT is shown in Figure 1. In Figure 1, the functional group test of the SBT is performed using Fourier transform infrared spectroscopy (Bruker Equinox 55, resolution 0.4 cm^{-1}). The absorption peaks of 3436.79 cm^{-1}, 2927.68 cm^{-1}, 1030.63 cm^{-1} and 766.30 cm^{-1} correspond to the tensile vibrations of hydroxyl (-OH), C-H, C-O and C-O-C, respectively.

Table 1. Chemical composition of raw materials/%.

Raw Materials	SiO_2	Al_2O_3	Fe_2O_3	CaO	MgO	K_2O	Na_2O	SO_3	Loss
Cement	18.55	3.95	3.41	65.32	1.01	0.72	0.18	2.78	2.88
FA	44.06	42.06	2.91	3.80	0.40	0.49	0.16	0.75	2.48
S95	33.39	11.89	0.63	41.51	8.82	0.53	0.67	/	0.28
MEA	3.87	1.03	0.88	1.98	89.37	0.88	/	0.06	2.38

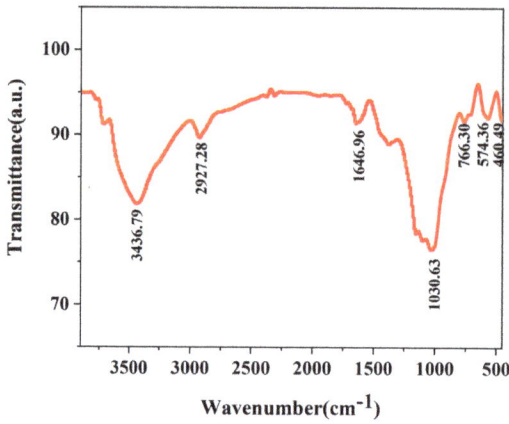

Figure 1. FTIR spectra of SBT with a resolution of 0.4 cm^{-1}.

2.2. Sample Preparation

In order to study the effect of hydration temperature rise inhibitor on concrete temperature rise and its mechanism, first of all, the concrete fit used in the experiment is shown in Table 2. The dosage of hydration temperature rise inhibitor was 0%, 0.5%, 1.0% and 1.5% of the total dosage of cementing material, respectively. Then, the specific experimental method is to stir these ingredients under dry conditions, and then add quantitative water to continue to stir evenly. The fully mixed slurry is placed in a mold of a specific size for the internal temperature rise of concrete and the 3 d, 7 d and 28 d compressive strength of concrete. In addition, without considering the addition of coarse and fine aggregates, after the raw materials are fully stirred, part of the slurry is put into a specific mold for a medium to 3 d, 7 d and 28 d age, which is used for micro-tests, such as hydration heat and thermogravimetry.

Table 2. Concrete proportioning (kg/m^3).

Number	Cement	FA	S95	MEA	Sand	Stones	Water	Water Reducing Agent	SBT
Blank	288	90	36	36	700	1100	144	12.6	0
0.5%SBT	288	90	36	36	700	1100	144	12.6	2.25
1.0%SBT	288	90	36	36	700	1100	144	12.6	4.5
1.5%SBT	288	90	36	36	700	1100	144	12.6	6.25

2.3. Methods

2.3.1. Macroscopic Experimental Methods

The macroscopic experimental methods in this paper include hydration heat test, concrete temperature rise test and mechanical test. The heat of hydration test adopts dissolution method, that is, isothermal calorimeter. Firstly, a certain amount of fresh cement paste was placed in the test tube, and then the data were monitored for 3 days. Finally, the hydration heat and cumulative heat release data of the cement paste were obtained.

The temperature rise test of concrete uses a thermometer to measure, the concrete is mixed into a fixed mold and then the thermometer is added to record the change in its internal temperature in real time. The insulation device uses an iron bucket with a bottom diameter of 300 mm and a height of 300 mm. The bottom and outside of the bucket are wrapped with 5 cm thick rubber and plastic insulated cotton. After the wrapping is completed, the whole bucket is put into the insulation bucket of the same size. The thermometer is inserted into the center of the concrete, and the thermometer connecting module records the internal temperature in real time. The mechanical properties were tested by universal pressure testing machine. In each age, 3 specimens were used to test the compressive strength, and their average values were recorded.

2.3.2. Microscopic Experimental Analysis

In this paper, the microscopic characterization techniques include XRD, TG, MIP, SEM, etc. At the end of the mechanical test, appropriate amount of the broken sample was soaked in anhydrous ethanol to stop hydration for a week, and then dried in vacuum at 80 °C for 2 days. After taking out the sample, it was used for SEM. The powder sample is then prepared, and the sample is ground into powder and passed through a 200-mesh screen to prepare the sample for XRD and TG analysis. The phase composition of the sample was characterized by XRD and TG. XRD analysis was used to observe the difference in the composition of crystalline mineral phase in the hydration products. The quantitative analysis of $Mg(OH)_2$ in cement slurry was performed by TG analysis. Obtained by STA 409PC Luxx at 10 °C/min from 30 °C to 1000 °C in a nitrogen atmosphere. The pore structure was characterized by mercury porosity method in the pressure range of 10–100 kPa, corresponding to pore size range of 1.7–300 nm. The microstructure of the concrete was characterized by SEM, in which the dry powder required prior spraying of gold, and scanned using Merlin Compact scanning electron microscopy.

3. Results and Discussion

3.1. Macroscopic Properties

3.1.1. Mechanical Property

As shown in Figure 2a, the compressive strength of concrete increases gradually with the increase in age. In the concrete mixed with SBT, the compressive strength decreases with the increase in the amount of SBT. In the first 3 days, it decreased significantly with the increase in variables. However, the reduction degree of compressive strength at 28 days decreased compared with that at 3 days. For example, the compressive strength of 0.5%SBT1 and 1.5%SBT1 was 14.4% and 37.2% lower than blank at 3 d, and 4.0% and 11.6% lower than blank at 28 d, respectively. The reason may be that SBT hindered the nucleation and growth of C-S-H in the early stage but did not affect the degree of hydration in the later stage [10].

Figure 2b shows the compressive strength of concrete specimens at different ages when they are placed in a curing box at variable temperature. The compressive strength of concrete increases under variable temperature conditions. For example, the compressive strength of 0.5%SBT2 and 1.5%SBT2 at 3 d increases by 24.1% and 20%, respectively, compared with that at normal temperature. This is because the higher the temperature, the higher the internal hydration rate and the higher the strength of the concrete.

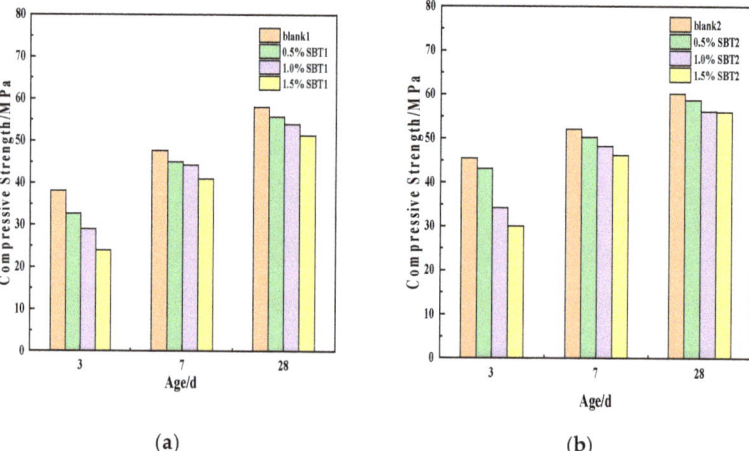

Figure 2. Compressive strength of concrete specimens at different ages. (**a**) is the compressive strength of concrete specimen under normal temperature curing; (**b**) is the compressive strength of concrete specimen under variable-temperature curing.

3.1.2. Temperature Rise Experiment

Figure 3 shows the internal temperature changes in concrete specimens after adding different amounts of SBT. As shown in the figure, with the increase in the amount of SBT, the temperature inside the concrete began to rise relatively late, and its peak temperature also decreased relatively. For example, the peak temperature of 0.5%SBT is 10.6 h later than blank, and the peak temperature is 4.5% lower. Further, 1.5%SBT is delayed for 32.3 h and the peak temperature is reduced by 7.5%. Equipment error is negligible. The reason is that the addition of SBT will hinder the nucleation and growth of C-H-S inside the concrete, thus reducing the hydration rate inside the concrete, resulting in a smaller internal temperature, resulting in a lower temperature rise inside the concrete. This is consistent with the conclusion above; that is, the temperature rise of concrete is slow, the peak temperature is low and the compressive strength of concrete specimens decreases.

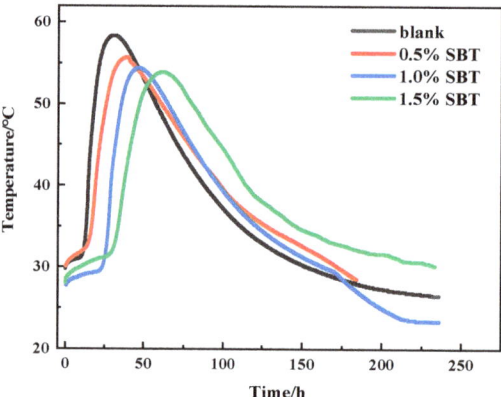

Figure 3. Internal temperature of concrete specimen.

3.1.3. Hydration Heat

Figure 4 shows the heat of hydration of the cement slurry in 3 d. According to previous conclusions, MEA can enhance the release of hydration heat, which is manifested as advanced induction period and increase in heat flow [18]. When SBT is added, the induction period and heat release peak are delayed relatively. With the increase in SBT incorporation, the early internal heat release decreased gradually. Therefore, for cement slurry, the addition of SBT will inhibit the internal hydration reaction, thus reducing the internal temperature rise and temperature peak. By extension, it can be concluded that the addition of hydration heat inhibitors can effectively reduce the temperature difference between inside and outside of concrete in the early stage, thus reducing the risk of concrete cracking.

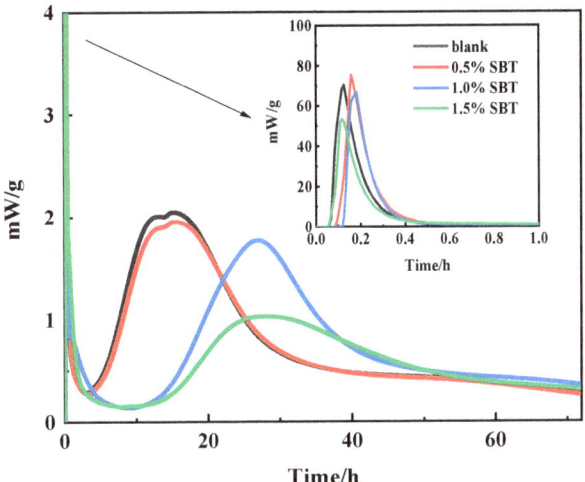

Figure 4. Heat map of hydration of cement paste.

3.2. Microstructural Characteristics

3.2.1. XRD

As can be seen from Figure 5, compared with the base group, the Ca(OH)$_2$ diffraction peak strength of the cement slurry mixed with different amounts of SBT is lower than that of the base group, the Ca(OH)$_2$ diffraction peak strength of the cement slurry mixed with 0.5%SBT hydration temperature rise inhibitor is the strongest among the three groups and the calcium hydroxide diffraction peak strength of the cement slurry mixed with 1.5%SBT is the lowest. The principle of cement hydration degree was tested by referring to CH quantitative test method. It is concluded that different amounts of SBT can effectively delay cement hydration, and, with the increase in its dosage, its inhibition effect will be better.

3.2.2. TG

Figure 6 shows the TG curves of the sample containing 1.0% SBT hydration heat inhibitor at 1, 3, 7 and 28 days. It corresponds to the dehydration decomposition of hydration products C-S-H, C-A-H, AFm, AFt, etc. [19]. Mg(OH)$_2$, Ca(OH)$_2$ and CaCO$_3$ will thermal decompose at peak absorption temperatures of 330–420 °C, 420–500 °C and 600–750 °C, respectively [20,21]. The relevant equations are shown in Equations (1)–(3). Table 3 shows the mass scores of blank and 1.0%SBT at different temperatures. Meanwhile, Mg(OH)$_2$ content can be calculated according to Equation (4) and Table 3, and the results are shown in Figure 7.

$$Mg(OH)_2 \rightarrow MgO + H_2O, \tag{1}$$

$$Ca(OH)_2 \rightarrow CaO + H_2O \tag{2}$$

$$CaCO_3 \rightarrow CaO + CO_2 \tag{3}$$

$$WMg(OH)_2 = \Delta W_{330-420°C} \frac{MMg(OH)_2}{MH_2O} \tag{4}$$

where $WMg(OH)_2$ is the mass fraction(%) of $Mg(OH)_2$; $MMg(OH)_2$ is the molar mass of $Mg(OH)_2$ (g/mol); $\Delta W_{330-420\ °C}$ represents the mass loss fraction(%) in the 330–420 °C temperature range.

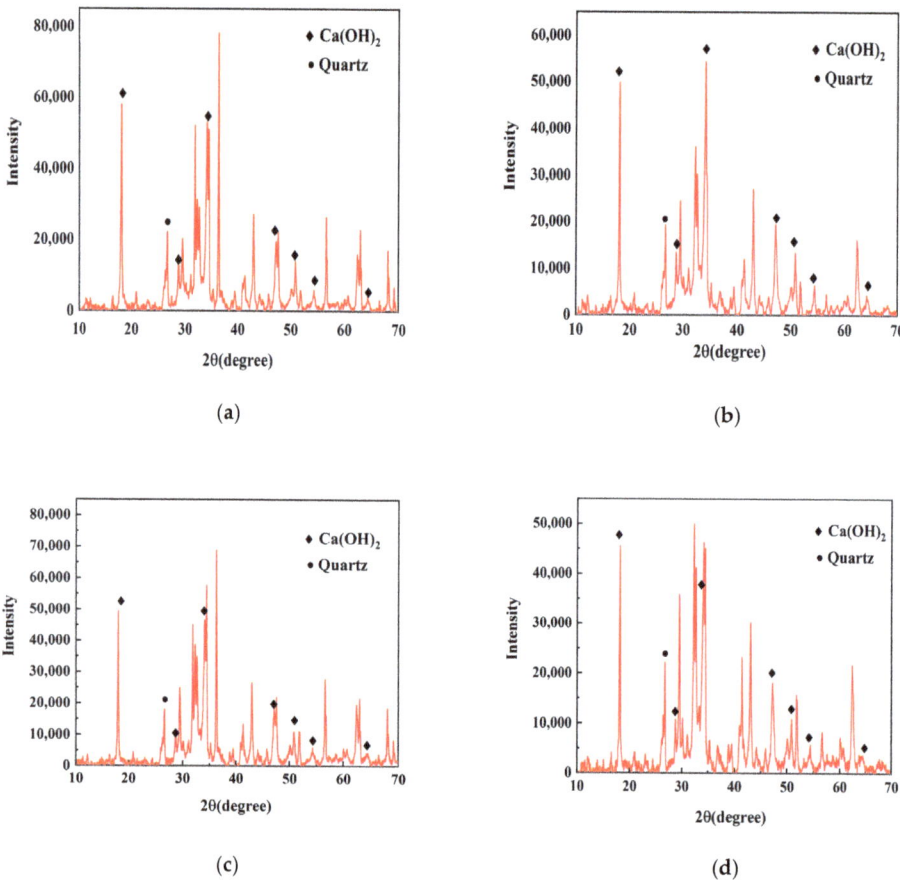

Figure 5. XRD pattern of cement slurry, where (**a**) is blank reference group; (**b**) the hydration temperature rise inhibitor group containing 0.5%SBT; (**c**) is the dosage of 1.0%SBT hydration temperature rise inhibitor group; (**d**) is the dosage of 1.5%SBT hydration temperature rise inhibitor group.

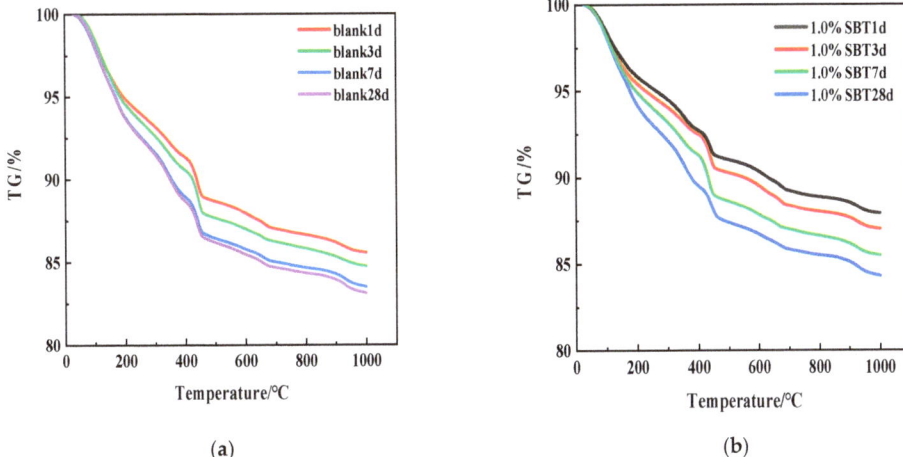

Figure 6. TG diagrams of cement slurry at 1 d, 3 d, 7 d and 28 d. (**a**) The TG curves of the blank group at 4 instars; (**b**) is the TG curve of 4 instars of hydration temperature rise inhibitor containing 1.0%SBT.

Table 3. The mass fraction of each sample at different temperatures.

Temperature (°C)		30	330	420	500	600	750
Blank	1 d	100	92.62	90.92	88.67	87.93	86.77
	3 d	100	91.97	90.15	87.63	86.96	85.98
	7 d	100	90.83	88.57	86.42	85.75	84.79
	28 d	100	90.68	88.25	86.16	85.44	84.46
1.0%SBT	1 d	100	92.44	92.44	91.01	90.35	89.22
	3 d	100	92.08	92.08	90.23	89.49	88.11
	7 d	100	90.63	90.63	88.58	87.81	86.96
	28 d	100	89.16	89.16	87.35	86.68	85.62

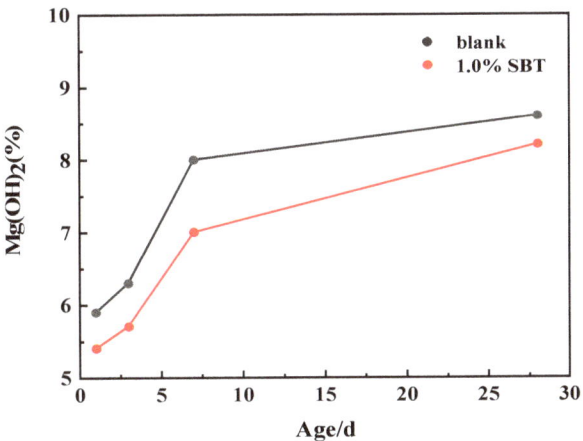

Figure 7. The degree of hydration of MgO at each age.

The degree of hydration of MgO can be discussed through quantitative analysis in the figure. As can be seen from Figure 7, Mg(OH)$_2$ of 1.0%SBT decreased in each age compared with blank, but the total amount still increased with age. The decrease in $Mg(OH)_2$ after the addition of SBT was due to the fact that SBT could adsorb on the surface of MEA and prevent the hydration of MgO, thus leading to the decrease in $Mg(OH)_2$. On the other hand, it is also possible that $Mg(OH)_2$ reacts with SiO_2 to form M-S-H and then $Ca(OH)_2$, which also relatively reduces the degree of hydration of $Mg(OH)_2$.

3.2.3. MIP

Another important parameter to reflect the microstructure is to observe the pore structure. It can be obtained from the aperture distribution and cumulative aperture distribution as shown in Figure 8. The pore size of blank group and 1.0%SBT group was mainly distributed between 10 nm and 100 nm. It can be seen from 3 d and 7 d data that the primary aperture of blank group and 1.0%SBT group at 7 d is larger than that of 3 d group. It can be seen from the figure that the main pore diameter of the sample after adding 1.0%SBT is larger than that of the blank group. In addition, it can be seen from Figure 8b that the porosity values from large to small are 1.0%SBT3d, blank3d, 1.0%SBT7d, blank7d. The microstructure of cement slurry without hydration heat inhibitor is denser and its corresponding compressive strength is relatively higher.

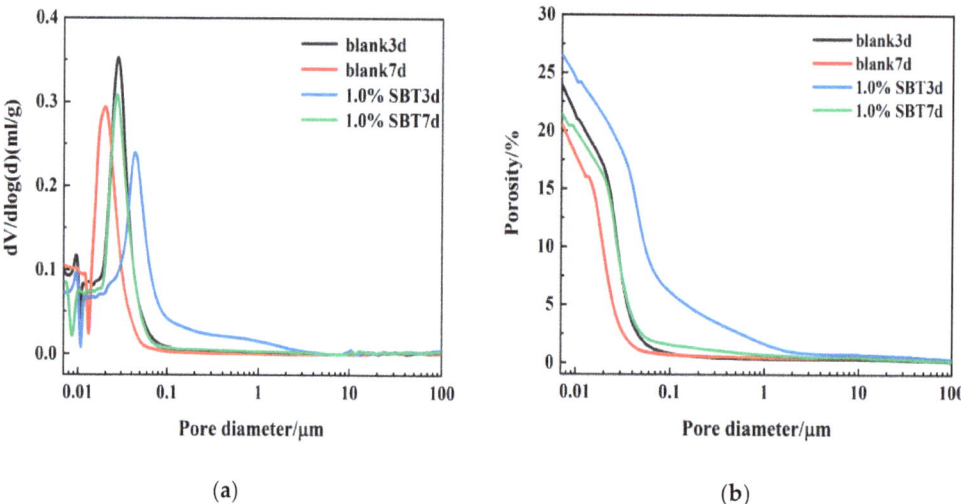

Figure 8. 3 d and 7 d mercury injection maps of blank cement paste group and 1.0%SBT. (**a**) is the aperture distribution diagram of each specimen; (**b**) is the cumulative aperture distribution diagram of each specimen.

3.2.4. SEM

In order to better understand the differences in mechanical properties between blank group and 1.0%SBT group in early stage, the concrete specimens were photographed by electron microscopy. Figure 9 shows SEM observations of concrete of corresponding age. On the whole, the concrete structure is relatively loose, and there are obvious voids and holes in the structure. As can be seen from Figure 9a, hydration products, such as acicular ettringite, hexagonal crystalline calcium hydroxide and flocculent calcium silicate hydrate, exist in the surface topography of concrete, and no other special substances are observed. In addition to the above substances, some white crystals circled as shown in the picture are also observed in Figure 9b, which are partially incomplete SBT residue and partially MgO, indicating that the hydration heat inhibitor delays the hydration process of cement and also reduces the degree of MgO hydration.

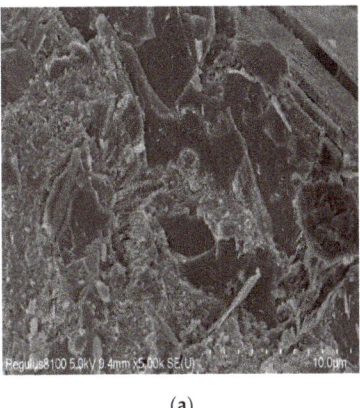 

(a) (b)

Figure 9. SEM image of 3 d concrete specimen. (**a**) SEM diagram of concrete specimen without water-added temperature rise inhibitor 3 d; (**b**) SEM diagram of concrete specimen doped with 1.0%SBT hydration temperature rise inhibitor.

4. Conclusions

In this paper, the effect and mechanism of hydration heat inhibitors on the temperature rise of concrete are studied systematically from two aspects of macroscopic property and microscopic structure. According to the above experimental results, the following conclusions can be drawn.

1. The compressive strength of concrete with hydration heat inhibitors decreased by 14.4–37.2% at 3 d. This indicates that the addition of heat of a hydration inhibitor can significantly reduce the early compressive strength of concrete. It was reduced by about 10% at 28 d to show that it had little effect on the concrete in the later stage.
2. It can be seen from the internal temperature rise diagram and hydration heat map of concrete that hydration heat inhibitor delays the time of concrete reaching the temperature peak by 10–30 h. Adding hydration heat inhibitors significantly reduces the heat flow and heat release rate of cement, prolongs the setting time of cement, inhibits the early hydration process of cement and improves the mechanical properties of cement in the later stage.
3. XRD and TG results show that the hydration heat inhibitor does not stop the hydration process of cement and has no effect on the type and morphology of hydration products. It can only delay the hydration process of cement and reduce the content of magnesium hydroxide and calcium hydroxide in hydration products. Hydration heat inhibitors can delay cement hydration.
4. The results of MIP experimental analysis and SEM image observation show that the addition of hydration heat inhibitors can increase the main pore size and porosity of cement, which is also the result that the hydration heat inhibitor delays the early hydration degree of cement. At the same time, the microstructure of concrete has not changed much. Specifically, the concrete structure is relatively loose, obvious voids and holes can be observed and only part of the MgO and hydration heat inhibitors are not fully reacted. This further indicates that the addition of heat of hydration inhibitor is not conducive to the microstructure densification of cement slurry and delays the degree of hydration of MgO.

Author Contributions: Conceptualization, M.T. and M.D.; data curation, T.L., P.L. and X.H.; writing—original draft preparation, T.L.; writing—review and editing, M.D. and Z.M. All authors have read and agreed to the published version of the manuscript.

Funding: This work was supported by the Open Fund project of National Laboratory for High Performance Civil Engineering Materials (2021CEM012) and the Priority Academic Program Development of Jiangsu Higher Education Institutions (PAPD).

Institutional Review Board Statement: Not applicable.

Informed Consent Statement: Not applicable.

Data Availability Statement: The data presented in this study are available on request from the corresponding author.

Acknowledgments: The authors gratefully acknowledge the assistance from Min Deng from NJTECH and the staff from State Key Laboratory of Materials—Oriented Chemical Engineering.

Conflicts of Interest: The authors declare no conflict of interest.

References

1. Hossain, A.B.; Weiss, J. Assessing Residual Stress Development and Stress Relaxation in Restrained Concrete Ring Specimens. *Cem. Concr. Compos.* **2004**, *26*, 531–540. [CrossRef]
2. Shi, N.N.; Ouyang, J.S.; Zhang, R.X.; Huang, D.H. Experimental Study on Early-Age Crack of Mass Concrete Under the Controlled Temperature History. *Adv. Mater. Sci. Eng.* **2014**, *2014*, 671795. [CrossRef]
3. Ha, J.H.; Jung, Y.S.; Cho, Y.G. Thermal Crack Control in Mass Concrete Structure Using an Automated Curing System. *Automat. Constr.* **2014**, *45*, 16–24. [CrossRef]
4. Li, M.; Xu, W.; Wang, Y.J.; Tian, Q.; Liu, J.P. Shrinkage Crack Inhibiting of Cast in Situ Tunnel Concrete by Double Regulation on Temperature and Deformation of Concrete at Early Age. *Constr. Build. Mater.* **2020**, *240*, 117834. [CrossRef]
5. Li, W.G.; Huang, Z.Y.; Hu, G.Q.; Duan, W.H.; Shah, S.P. Early-Age Shrinkage Development of Ultra-High-Performance Concrete Under Heat Curing Treatment. *Constr. Build. Mater.* **2017**, *131*, 767–774. [CrossRef]
6. An, G.; Park, J.; Cha, S.; Kim, J. Development of a Portable Device and Compensation Method for the Prediction of the Adiabatic Temperature Rise of Concrete. *Constr. Build. Mater.* **2016**, *102*, 640–647. [CrossRef]
7. Wang, Z.H.; Tao, L.; Liu, Y.; Jiang, Y.H. Temperature Control Measures and Temperature Stress of Mass Concrete during Construction Period in High-Altitude Regions. *Adv. Civ. Eng.* **2018**, *2018*, 9249382. [CrossRef]
8. Mbugua, R.; Salim, R.; Ndambuki, J. Effect of Gum Arabic karroo as a Water-Reducing Admixture in Concrete. *Materials* **2016**, *9*, 80. [CrossRef]
9. Poole, J.L.; Riding, K.A.; Juenger, M.; Folliard, K.J.; Schindler, A.K. Effect of Chemical Admixtures on Apparent Activation Energy of Cementitious Systems. *J. Mater. Civ. Eng.* **2011**, *23*, 1654–1661. [CrossRef]
10. Zhang, H.; Wang, W.B.; Li, Q.L.; Tian, Q.; Li, L.; Liu, J.P. A Starch-Based Admixture for Reduction of Hydration Heat in Cement Composites. *Constr. Build. Mater.* **2018**, *173*, 317–322. [CrossRef]
11. Mwaluwinga, S.; Ayano, T.; Sakata, K. Influence of Urea in Concrete. *Cem. Concr. Res.* **1997**, *27*, 733–745. [CrossRef]
12. Makul, N.; Sua-iam, G. Effect of Granular Urea on the Properties of Self-Consolidating Concrete Incorporating Untreated Rice Husk Ash: Flowability, Compressive Strength and Temperature Rise. *Constr. Build. Mater.* **2018**, *162*, 489–502. [CrossRef]
13. Zhang, H.; Li, L.; Wang, W.B.; Liu, J.P. Effect of Temperature Rising Inhibitor on Expansion Behavior of Cement Paste Containing Expansive Agent. *Constr. Build. Mater.* **2019**, *199*, 234–243. [CrossRef]
14. Zhang, H.; Liu, X.; Feng, P.; Li, L.; Wang, W.B. Influence of Temperature Rising Inhibitor on Nucleation and Growth Process During Cement Hydration. *Thermochim. Acta* **2019**, *681*, 178403. [CrossRef]
15. Yan, Y.; Ouzia, A.; Yu, C.; Liu, J.P.; Scrivener, K.L. Effect of a Novel Starch-Based Temperature Rise Inhibitor on Cement Hydration and Microstructure Development. *Cem. Concr. Res.* **2020**, *129*, 105961. [CrossRef]
16. Zhang, H.; Li, L.; Feng, P.; Wang, W.B.; Tian, Q.; Liu, J.P. Impact of Temperature Rising Inhibitor on Hydration Kinetics of Cement Paste and its Mechanism. *Cem. Concr. Compos.* **2018**, *93*, 289–300. [CrossRef]
17. Kim, H. Urea Additives for Reduction of Hydration Heat in Cement Composites. *Constr. Build. Mater.* **2017**, *156*, 790–798. [CrossRef]
18. He, J.; Zheng, W.H.; Bai, W.B.; Hu, T.T.; He, J.H.; Song, X.F. Effect of Reactive MgO On Hydration and Properties of Alkali-Activated Slag Pastes with Different Activators. *Constr. Build. Mater.* **2021**, *271*, 121608. [CrossRef]
19. Alarcon-Ruiz, L.; Platret, G.; Massieu, E.; Ehrlacher, A. The Use of Thermal Analysis in Assessing the Effect of Temperature on a Cement Paste. *Cem. Concr. Res.* **2005**, *35*, 609–613. [CrossRef]

20. Zhao, Z.G.; Qu, X.L.; Li, J.H. Microstructure and Properties of Fly Ash/Cement-Based Pastes Activated with MgO and CaO Under Hydrothermal Conditions. *Cem. Concr. Compos.* **2020**, *114*, 103739. [CrossRef]
21. Bahafid, S.; Ghabezloo, S.; Duc, M.; Faure, P.; Sulem, J. Effect of the Hydration Temperature on the Microstructure of Class G Cement: C-S-H Composition and Density. *Cem. Concr. Res.* **2017**, *95*, 270–281. [CrossRef]

Disclaimer/Publisher's Note: The statements, opinions and data contained in all publications are solely those of the individual author(s) and contributor(s) and not of MDPI and/or the editor(s). MDPI and/or the editor(s) disclaim responsibility for any injury to people or property resulting from any ideas, methods, instructions or products referred to in the content.

Article

Effect of Working Temperature Conditions on the Autogenous Deformation of High-Performance Concrete Mixed with MgO Expansive Agent

Zhe Cao [1], Zhongyang Mao [1], Jiale Gong [1], Xiaojun Huang [1] and Min Deng [1,2,*]

[1] College of Materials Science and Engineering, Nanjing Tech University, Nanjing 211816, China; 202061103063@njtech.edu.cn (Z.C.)

[2] State Key Laboratory of Material-Oriented Chemical Engineering, Nanjing Tech University, Nanjing 211800, China

* Correspondence: dengmin@njtech.edu.cn; Tel.: +86-136-0518-4865

Abstract: Currently, mass concrete is increasingly utilized in various engineering projects that demand high physical properties of concrete. The water-cement ratio of mass concrete is comparatively smaller than that of the concrete used in dam engineering. However, the occurrence of severe cracking in mass concrete has been reported in numerous engineering applications. To address this issue, the incorporation of MgO expansive agent (MEA) in concrete has been widely recognized as an effective method to prevent mass concrete from cracking. In this research, three distinct temperature conditions were established based on the temperature elevation of mass concrete in practical engineering scenarios. To replicate the temperature increase under operational conditions, a device was fabricated that employed a stainless-steel barrel as the container for concrete, which was enveloped with insulation cotton for thermal insulation purposes. Three different MEA dosages were used during the pouring of concrete, and sine strain gauges were placed within the concrete to gauge the resulting strain. The hydration level of MEA was studied using thermogravimetric analysis (TG) to calculate the degree of hydration. The findings demonstrate that temperature has a significant impact on the performance of MEA; a higher temperature results in more complete hydration of MEA. The design of the three temperature conditions revealed that when the peak temperature exceeded 60 °C in two cases, the addition of 6% MEA was sufficient to fully compensate for the early shrinkage of concrete. Moreover, in instances where the peak temperature exceeded 60 °C, the impact of temperature on accelerating MEA hydration was more noticeable.

Keywords: MgO expansive agent; autogenous shrinkage; temperature

1. Introduction

The utilization of mass concrete has become increasingly prevalent in various civil engineering projects such as high-rise buildings, tunnels, and box girders, due to the upsurge in demand for such projects. Concrete is a fragile material with a lower tensile strength compared to its compressive strength, and mass concrete possesses the attributes of high total heat release and a large volume. The substantial size of the concrete impedes the timely dissipation of the heat generated from hydration, causing a rapid and considerable temperature increase in the mass concrete, with a significant temperature differential between the interior and exterior. Previously, dams utilized concrete with a high water-cement ratio and a relatively small amount of cement, leading to low heat of cement hydration and internal temperatures below 50 °C. However, in contemporary large-scale projects like retaining walls and bearing platforms, the utilization of mass concrete necessitates a smaller water-cement ratio, resulting in high heat of hydration and central temperatures above 70 °C, generating tensile stress that surpasses the concrete's tensile limit, ultimately leading to cracking [1,2].

Citation: Cao, Z.; Mao, Z.; Gong, J.; Huang, X.; Deng, M. Effect of Working Temperature Conditions on the Autogenous Deformation of High-Performance Concrete Mixed with MgO Expansive Agent. *Materials* 2023, *16*, 3006. https://doi.org/10.3390/ma16083006

Academic Editor: Alessandro P. Fantilli

Received: 13 March 2023
Revised: 7 April 2023
Accepted: 8 April 2023
Published: 10 April 2023

Copyright: © 2023 by the authors. Licensee MDPI, Basel, Switzerland. This article is an open access article distributed under the terms and conditions of the Creative Commons Attribution (CC BY) license (https://creativecommons.org/licenses/by/4.0/).

In the context of mass concrete with a low water-cement ratio, autogenous shrinkage is a crucial factor that contributes to cracking. In engineering, autogenous shrinkage is particularly pronounced in high-strength concrete, self-consolidating concrete, and mass concrete. Research indicates that high-strength concrete with a water-cement ratio of less than 0.3 can experience an autogenous shrinkage of 300–400 με and that the magnitude of autogenous shrinkage increases with the amount of cement per unit volume [3]. According to Erika Holt, the impact of the water-cement ratio on the autogenous shrinkage of concrete is generally insignificant when the ratio is greater than 0.4, but significant when it is less than 0.4 [4]. The autogenous shrinkage of concrete can be influenced by various factors such as the water-cement ratio, type of cement used, and mineral admixtures. According to Li's investigation, the self-shrinkage of concrete without mineral admixtures is higher than that of concrete containing mineral admixtures. The self-shrinkage of mineral admixture concrete stabilizes after seven days, while non-admixture concrete continues to exhibit high levels of self-shrinkage for up to three weeks. While the self-shrinkage of pure cement is greater than that of mineral admixture concrete, it generates lower stress, and the cracking time of non-mineral admixture concrete is delayed [5]. Similarly, Lee's research reveals that the self-shrinkage of slag concrete is greater than that of non-slag concrete, and the degree of self-shrinkage is directly proportional to the amount of slag present. This can be attributed to the fact that the particle size of slag concrete is finer than that of cement, resulting in a greater surface area, and denser pore structure following hydration. The chemical shrinkage value of concrete containing slag is higher compared to pure cement concrete [6]. A study conducted by Aveline revealed that while the autogenous shrinkage of slag cement concrete is higher, the time to cracking is longer than Portland cement concrete. However, under restrained conditions, slag cement concrete is more prone to cracking [7]. Laurent Barcelo's investigation into the influence of SO_3 content in cement on the dry shrinkage of concrete demonstrated that the dry shrinkage of concrete reduces as the SO_3 content increases, as long as it does not surpass 3.1%. On the other hand, if it exceeds 3.1%, the trend is reversed [8]. Currently, there are several methods available for early concrete shrinkage measurement, including the volume method, length method, embedded sensor method, and other methods. Every approach has its own merits and demerits. When it comes to concrete, the volume method cannot be utilized due to the likelihood of aggregates causing harm to rubber bags. On the other hand, the length method is intricate and has a high probability of human errors. The most appropriate approach for measuring the initial deformation of concrete is through embedded sensors, although sine strain gauges are costly and non-reusable [9–11]. In this research, to precisely evaluate the early autogenous deformation of concrete, the embedded sensors are employed to determine the initial shrinkage of concrete.

Expansive agents are commonly utilized in dam engineering to counteract the notable shrinkage of mass concrete and to forestall cracking. The MgO expansive agent (MEA) is a particularly promising candidate due to its superior expansion performance and the adjustability of its expansion behavior [12,13]. In recent years, the application of MEA in practical engineering projects in China [14–17] has yielded systematic achievements in both the understanding of the expansion mechanism and the control of expansion amount [18,19] leading to the establishment of a comprehensive compensating system for mass concrete. The expansion performance of MEA is influenced by various factors, including the water-cement ratio, mineral admixtures, and curing temperature, with the latter being a significant determinant of MEA's performance. Extensive research has confirmed that temperature exerts a profound impact on the reactivity of MEA. Specifically, a higher curing temperature accelerates the hydration rate of MEA, as evidenced by several studies [20–22]. Liu posits that the hydration process of mildly calcined MgO, akin to cement hydration, comprises five stages, including the initial, induction, acceleration reaction, deceleration reaction, and stable stages. The influence of temperature on the hydration reaction of MgO is significant, leading to pronounced temperature sensitivity in the expansion behavior of the MEA-cement slurry [23]. However, the hydration process of MEA in cement differs from that in pure form because of the competition for water between MEA and cement, leading to a significantly slower hydration rate of MEA in cement [24].

Given the high temperature sensitivity of both MEA and cement, investigating the impact of temperature on the deformation of MEA concrete is a crucial endeavor. There have been several investigations concerning the impact of curing temperature on the performance of MEA concrete. However, these studies were carried out mostly under conditions of high water-cement ratios and constant curing temperatures, which are inconsistent with the conditions encountered in large-scale vertical walls and load-bearing platforms where the temperature of the concrete can significantly increase due to the heat of hydration. Such engineering projects often require the use of a large number of cementitious materials and a relatively high water-cement ratio for dam concrete to achieve high concrete strength. Consequently, the central temperature of the concrete may reach over 70 °C, and the temperature is not constant. The use of constant temperature conditions in studying MEA concrete does not reflect the actual conditions encountered in engineering projects. To address this limitation, this study employs an insulation tool that can simulate the temperature change process of concrete under working conditions. By incorporating temperature and sine-type strain gauges into the concrete and monitoring the temperature and strain development of the concrete, the study aims to simulate the concrete strain in engineering structures and investigate the deformation performance of MEA concrete under different engineering temperatures.

2. Materials and Methods

2.1. Materials

In this study, ordinary Portland cement (OPC, P. II52.5) manufactured by Onoda Cement in Nanjing, China was employed. Fly ash (FA) was procured locally in Nanjing, China, while granulated blast furnace slag powder, designated as S95 mineral powder, was provided by Jiangsu Huailong Building Materials Co, Nanjing, China. The fine aggregate (fineness modulus of 2.7) had a well-graded size distribution and a well-rounded shape. Three types of coarse aggregate with the size grading of large size stone (20–25 mm), middle size stone (10–20 mm), and small size stone (5–10 mm) were used in the ratio of 2:5:3. The MEA used in the study was supplied by Jiangsu Subot, Nanjing, China, and was tested to have reaction values of 300 s using the citric acid method. The chemical composition, particle size, and specific surface area of the various adhesive materials used are presented in Table 1. It is noteworthy that the MEA had an average particle size of 22.76 μm. The particle size distribution curves of the materials are shown in Figure 1. The high-efficiency water-reducing agent (SP) used was polycarboxylate superplasticizer and was supplied by Jiangsu Subot, China. The water reduction rate of SP was determined to be 30%.

Table 1. Chemical compositions and physical properties of the raw materials.

Chemical Composition	OPC	FA	SLAG	MEA
CaO (%)	65.32	3.8	44.06	1.88
SiO_2 (%)	18.55	44.06	42.06	4.07
Al_2O_3 (%)	3.95	42.06	3.8	0.86
Fe_2O_3 (%)	3.41	2.91	0.57	0.57
MgO (%)	1.01	0.4	2.91	90.45
K_2O (%)	0.72	0.49	0.49	/
Na_2O (%)	0.18	/	0.75	/
SO_3 (%)	2.78	0.75	/	/
LOI (%)	2.88	2.48	2.28	1.53
physical properties				
specific surface area (m^2/g)	1.24	0.973	1.46	1.28
D50 (μm)	16.083	24.438	12.428	22.762

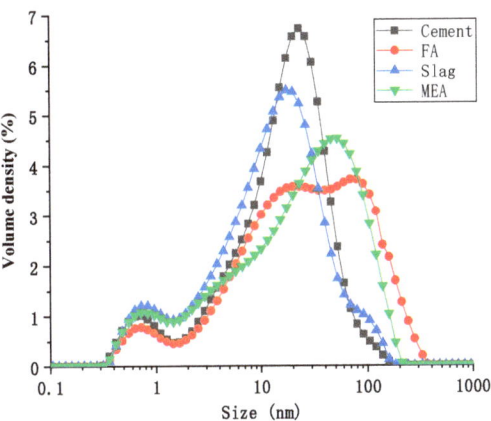

Figure 1. The particle size distribution curves of the materials.

2.2. Tools for Simulating the Temperature of Working Conditions

This study aimed to develop tools that could simulate working conditions' temperatures. The tools consisted of a stainless-steel barrel used as a container for the concrete, which was wrapped with insulation cotton around its exterior. The tool's ability to simulate various temperature changes was achieved by adjusting the volume of the concrete and the thickness of the insulation layer and by selecting different insulation materials. Three tools were selected from the fabricated tools to match the actual temperature rise curve of the poured concrete. The simulation tools are visually represented in Figures 2 and 3. The chosen insulation material is rubber and plastic thermal insulation cotton, with a thermal conductivity coefficient of 0.0302 W/m·K. The concrete volume and insulation layer thickness selected for the tools are shown in Table 2. A comparison of the temperature rise of concrete in the simulation tools with the actual wall temperature is presented in Figure 4. Temperature measurement was carried out by inserting a temperature sensor into the concrete. Temperature conditions were categorized based on the temperature peak, with the curve having the lowest temperature peak referred to as temperature condition A, the curve with a moderate temperature peak referred to as temperature condition B, and the curve with the highest temperature peak referred to as temperature condition C.

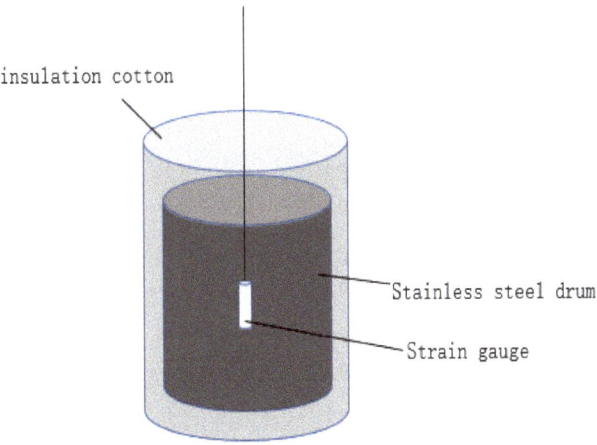

Figure 2. Schematic diagram of the simulation tool.

Figure 3. The simulation tools.

Table 2. The concrete volume and the thickness of insulation layer selected for the tools.

	Tool A	Tool B	Tool C
Concrete Vlome/m^3	0.02	0.05	0.07
Thickness of insulation layer/cm	5	10	15

To ensure that the simulation tools can effectively mimic the actual engineering structures, this study referred to the temperature-rise curve of a large thin-wall structure in a real project. The wall has a total length of 24 m, a height of 3.4 m, and a thickness of 1.1 m. The temperature rise was measured by using built-in strain gauges at the center, surface, and bottom of the wall, and the temperature rise of the wall is shown in Figure 5. The temperature rise at the center of the wall was the highest, reaching a peak temperature of 75.4 °C after 34 h of pouring; followed by the surface temperature rise, which reached a peak temperature of 67.8 °C after 22 h of pouring; the lowest temperature was at the bottom of the wall, reaching a peak temperature of 57.6 °C after 22 h of pouring.

2.3. Preparation of Concrete

Table 3 displays the mixture proportions of High-Performance Concrete (HPC) containing varying levels of MEA. The incorporation of MEA was achieved via internal admixture, at proportions of 0%, 6%, 8%, and 10% of the total cementitious material. The water-binder ratio for the concrete was set at 0.32. The manufacturing process involved weighing the raw materials in accordance with the mix proportions specified in Table 3, followed by the mixing of aggregates and binders in a mixer for one minute. Subsequently, water and a water-reducing agent were added to the mixer and stirred for five minutes. The dosage of polycarboxylate is 2.5% of the weight of the cementitious material. Following the mixing process, the concrete was poured into the simulation tools described in Section 2.2 for curing, in a completely dry state. The temperature rise is shown in Figure 5.

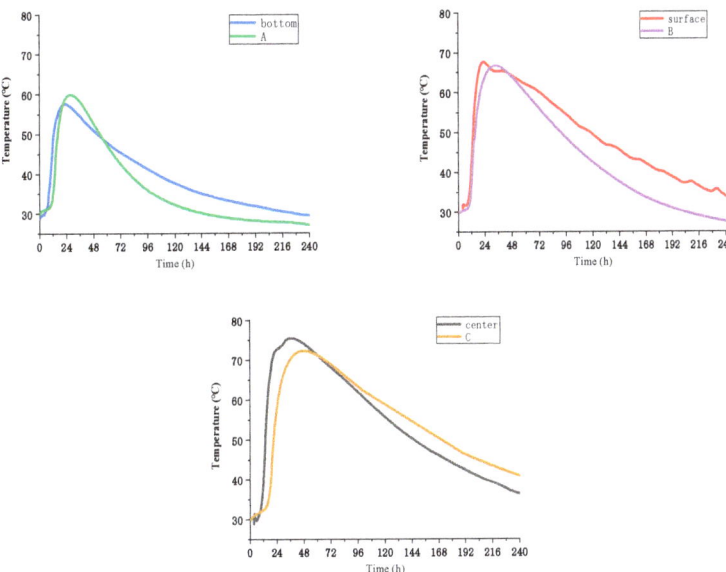

Figure 4. Comparison of the temperature rise of concrete in the simulation tools with the actual wall temperature.

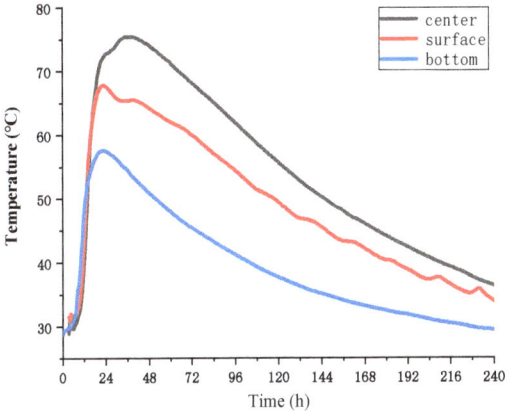

Figure 5. Temperature rise of the large thin wall.

Table 3. Mix proportions of HPC matrix (kg/m³).

OPC	FA	Slag	MEA	Fine Aggregate	Coarse Aggregate			Water
					5–10 mm	10–20 mm	20–30 mm	
310	90	40	50	136	220	550	330	157
320	90	40	40	136	220	550	330	157
330	90	40	30	136	220	550	330	157
360	90	40	0	136	220	550	330	157

2.4. Test Methods

2.4.1. Mechanical Properties

Based on the Chinese standard GB/T 50081-2019 [25], following the completion of the HPC mixing, the composite material was introduced into a mold with dimensions of 150 mm × 150 mm × 150 mm. Subsequently, the resulting specimens were transferred into a high-temperature maintenance box and subjected to incremental heating to replicate diverse temperature rise conditions for the compressive strength assessment. The compressive strength of the HPC samples for each mixture proportion at 3 d and 7 d was determined using a 2000 kN testing machine. To ensure the precision of the outcomes, three specimens were tested in each experimental group, and the mean compressive strength value of the three specimens was adopted as the final result.

2.4.2. Autogenous Deformation

Temperature and Autogenous deformation of concrete at early ages were measured by using the tools described in Section 2.2, and three simulations were selected to simulate three temperature rise conditions. The tools were placed in a room with a constant temperature of 20 °C. The concrete was mixed with a mixer. After mixing, the concrete was vibrated with a vibrating table and then poured into the simulation tool with a built-in strain gauge to detect the temperature rise and strain of the concrete. The temperature and strain of the concrete changed rapidly in the early stage, so to be able to track them continuously, the strain gauges were tracked by an automatic measuring instrument manufactured by Ge Nan Industrial Co., Nanjing, China. The collection interval was 5 min. The total monitoring time was 14 days.

2.4.3. Thermal Analysis

To evaluate the hydration extent of MgO, the thermogravimetric technique was employed on HPC mixtures lacking aggregate at the 3 d and 7 d stages in accordance with the Chinese standard GB/T 22314-2008 [26]. Following curing, the cement pastes were sectioned into small fragments, immersed in anhydrous ethanol for a minimum of 24 h to stop the hydration, and subsequently dried in a blast oven maintained at 60 °C for 24 h. The resultant small fragments were pulverized into a fine powder and filtered using a 0.075 μm square hole sieve. The powder was then subjected to heating from 30 °C to 1000 °C at a temperature ramp rate of 20 °C/min under an N_2 atmosphere via the use of a TAQ600 synchronous thermal analyzer.

2.4.4. SEM Morphology

To examine the microstructure of cement hydration products containing MEA, small portions of the cement pastes were sectioned, soaked in anhydrous ethanol for a minimum of 24 h to stop the hydration, and subsequently dried at 60 °C for 24 h. The resulting samples were then coated with gold and visualized using the Hitachi Regulus 8100 scanning electron microscope.

3. Results and Discussion

3.1. Mechanical Properties

Figure 6 depicts the compressive strength of high-performance concrete (HPC) containing MEA at 3 day, 7 day, and 14 day intervals under three different curing temperatures, A, B, and C. The results indicate that an increase in MEA content causes a gradual decrease in the compressive strength of HPC. Specifically, when cured under temperature A, the compressive strength at 14 days decreased by 1.2 MPa, 4.7 MPa, and 7.9 MPa for HPC samples containing 6%, 8%, and 10% MEA content, respectively, in comparison to samples without MEA. Similarly, when cured under temperature B, the compressive strength at 14 days decreased by 2.3 MPa, 6.3 MPa, and 7.7 MPa for HPC samples containing 6%, 8%, and 10% MEA content, respectively, in comparison to samples without MEA. Additionally, when cured under temperature C, the compressive strength at 14 days decreased by 1.5 MPa, 4.2 MPa, and 10.1 MPa for HPC samples containing 6%, 8%, and 10% MEA content, respectively, in comparison to samples without MEA. These findings align with previous studies that indicate the addition of MEA has a detrimental effect on concrete compressive strength because MEA partially replaces cement in the binding material., leading to a reduction of calcium hydroxide and other essential materials that contribute to strength [27–30]. Furthermore, the weakened mechanical properties of the MEA HPC composite system could be attributed to the relatively diminutive crystal phase of the hydrated MgO products, shown in Equation (1), which possess inferior strength when compared to the cement hydration products. This fact implies that the strength of the system is significantly affected [31].

$$Mg^{2+} + 2\,OH^- \rightarrow Mg(OH)_2 \qquad (1)$$

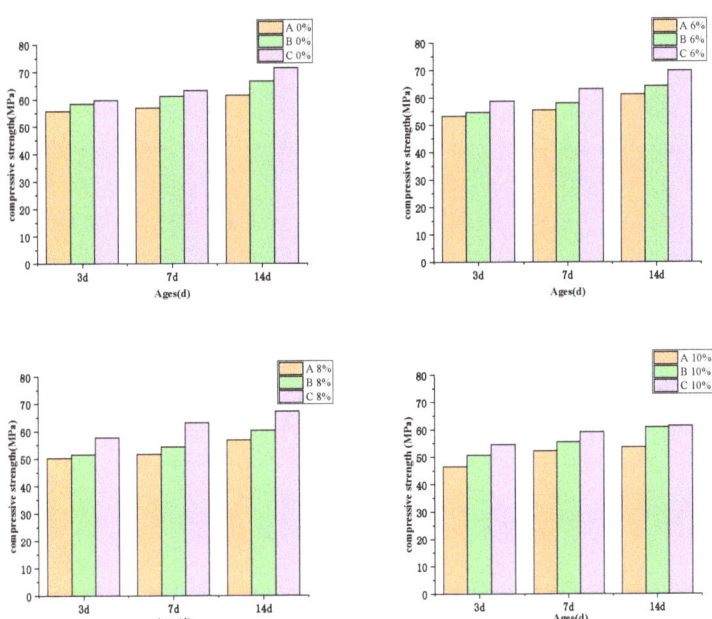

Figure 6. Effects of temperature conditions on the compressive strength of HPC mixed with MEA.

It is worth noting that the decline in strength observed at temperature C is relatively less pronounced compared to temperatures A and B, which could be attributed to the swift hydration of cement at high temperatures, which offsets the deleterious effects of MEA on the compressive strength of concrete.

3.2. Autogenous Deformation

Figure 7 presents the autogenous shrinkage behavior of concrete with varying amounts of MEA content over a 14 day period, while subjected to similar temperature conditions. Concrete samples without MEA exhibited significant shrinkage at all three temperature conditions, with respective shrinkage values of 218.3 με, 111.6 με, and 145.7 με at temperatures A, B, and C. The samples demonstrated faster shrinkage rates prior to day 4, and the trend of shrinkage became relatively stable after day 4. As evident, higher MEA content resulted in greater expansion of concrete. Specifically, under temperature condition B, concrete samples containing 6%, 8%, and 10% MEA exhibited expansion rates of 159.7 με, 361.0 με, and 57.2 με, respectively. Similarly, under temperature condition C, concrete samples with MEA contents of 6%, 8%, and 10% exhibited expansion rates of 257.1 με,

657.9 µε, and 713.3 µε, respectively. It is important to note that at temperature condition A, adding 6% of MEA was not sufficient to fully offset the concrete shrinkage after 14 days. However, adding 8% of MEA at temperature condition A exhibited an adequate mitigating effect on shrinkage, while adding 6% of MEA proved effective at reducing shrinkage at temperature conditions B and C.

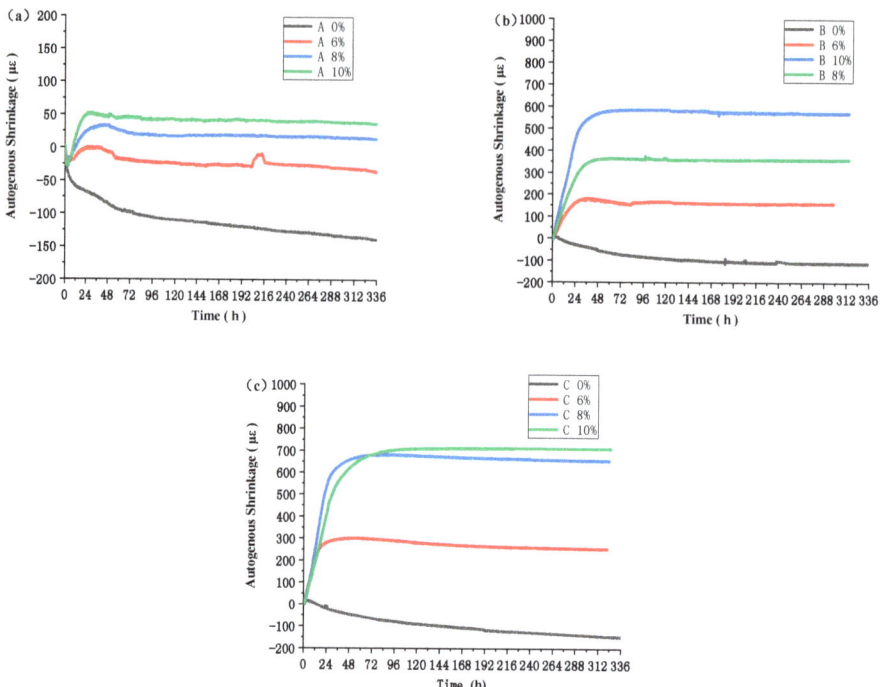

Figure 7. Effect of MEA content on early autogenous deformation of concrete at different temperature conditions. (**a**) Autogenous deformation at temperature A; (**b**) Autogenous deformation at temperature B; (**c**) Autogenous deformation at temperature C.

Figure 8 displays the impact of varying temperature conditions on autogenous deformation while maintaining a constant MEA content. The findings reveal that the performance of MEA improves with elevated concrete temperatures. In the case of concrete with 6% MEA content, it experiences a shrinkage of 37.37 µε under temperature condition A, while it undergoes expansion under temperature conditions B and C, with expansion values of 159.7 µε and 257.1 µε, respectively. For concrete with 8% MEA content, under temperature conditions B and C, the expansion values increase by 55.7% and 183.8%, respectively, compared to condition A. Likewise, under temperature conditions B and C, the expansion values of concrete with 10% MEA content increase by 128.33% and 184.60%, respectively, compared to condition A. The effect of temperature on the performance of MEA has been established in previous studies, which have demonstrated that MEA undergoes a hydration process similar to cement, and its expansion value increases with higher curing temperatures [32].

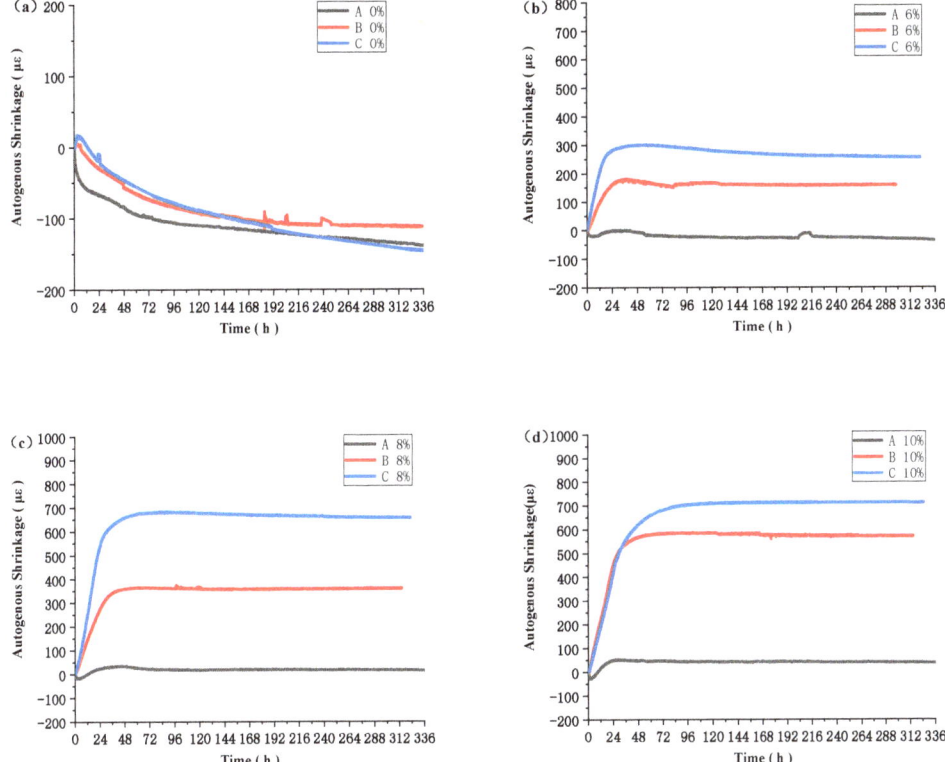

Figure 8. Effect of temperature on autogenous deformation of MEA concrete at early stage. (**a**) autogenous deformation, containing 0% MEA. (**b**) autogenous deformation, containing 6% MEA. (**c**) autogenous deformation, containing 8% MEA. (**d**) autogenous deformation, containing 10% MEA.

3.3. Hydration Degree of MgO

Figure 9 shows the TG/DTG curves of cement specimens mixed with MEA for 3 d and 7 d. As shown in Figure 9b,d, there are three different weight loss peaks on the DTG curve at 310–400 °C, 400–460 °C, and 650–750 °C, which correspond to the decomposition of $Mg(OH)_2$, $Ca(OH)_2$, and $CaCO_3$, respectively. The degree of hydration of $Mg(OH)_2$ and MgO(H MgO) was calculated using Equations (2) and (3), respectively, according to the literature [33,34].

$$\text{Mass Mg(OH)}_2 = \frac{58 \times \text{Mass loss}(310\ ^\circ\text{C} \sim 400\ ^\circ\text{C})}{18} \quad (2)$$

$$H_{MgO} = \frac{40 \times \text{Mass loss}(310\ ^\circ\text{C} \sim 400\ ^\circ\text{C})}{n \times 18 \times [1 - \text{Mass loss}(950\ ^\circ\text{C})]} \quad (3)$$

where the Mass $Mg(OH)_2$ represents the content of $Mg(OH)_2$ formed in the hydrated cement pastes; the n represents the amounts of MEA in cement paste.

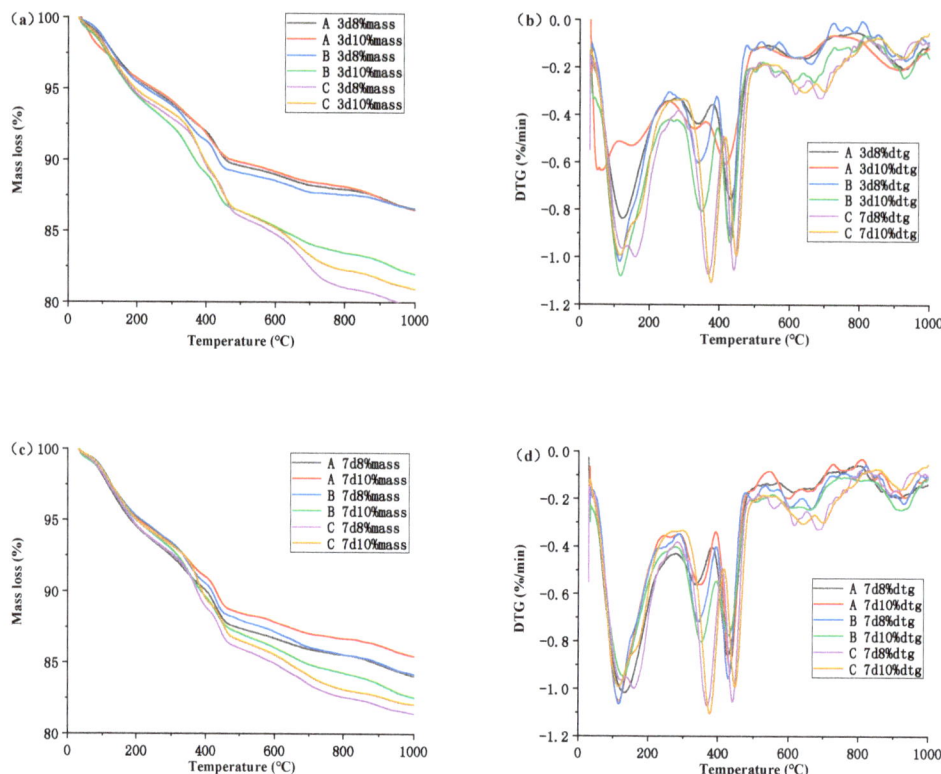

Figure 9. TG/DTG curves of the cement paste containing 8 wt% and 10 wt% MEA under different temperatures at 3 d and 7 d. (**a**) TG curves, at 3 d; (**b**) DTG curves, at 3 d; (**c**) TG curves, at 7 d; (**d**) DTG curves, at 7 d.

Table 4 illustrates the concentration of $Mg(OH)_2$ and the extent of MgO hydration in cement paste that incorporates MEA. It can be observed that the addition of MEA results in an increase in the content of $Mg(OH)_2$ in the samples, and a higher dosage of MEA corresponds to a greater degree of MgO hydration. For instance, under temperature condition A, the degree of MgO hydration after seven days was 32.41% and 38.19% when 8% and 10% of MEA were added to the cement, respectively. By comparing the degree of MgO hydration at three different temperature conditions, it can be inferred that the extent of MgO hydration escalates with an increase in the curing temperature. When the MEA concentration is 8%, the degree of MgO hydration at temperatures A, B, and C are 32.41%, 65.71%, and 84.34%, respectively. When the MEA concentration is 10%, the degree of MgO hydration at temperatures A, B, and C are 38.19%, 74.98%, and 84.40%, respectively. Evidently, higher curing temperatures facilitate the hydration of MgO. The degree of MEA hydration at temperature A is only 30%, whereas, at temperature B, it surpasses 60%. After seven days of hydration at temperature C, the degree of MEA hydration is around 80%. This implies that MEA hydrates expeditiously when the temperature exceeds 60 °C, and the rate of hydration does not increase proportionally with the temperature. The enhancing effect is relatively feeble before 60 °C.

Table 4. Estimated quantity of Mg(OH)$_2$ and hydration degree of MgO in cement pastes containing MEA.

	A 8%		A 10%		B 8%		B 10%		C 8%		C 10%	
	3 d	7 d	3 d	7 d	3 d	7 d	3 d	7 d	3 d	7 d	3 d	7 d
Mass loss at 320–400 °C/wt%	0.84	0.89	1.07	0.87	1.25	1.80	2.06	2.01	1.68	2.23	2.09	2.25
Mass Mg(OH)$_2$/%	2.69	2.86	2.80	3.46	4.04	5.80	6.63	6.49	5.41	7.19	6.74	7.25
H MgO	30.44	32.41	31.29	38.19	44.54	65.71	67.17	74.98	74.76	84.34	79.48	84.40

3.4. Morphology

Figure 10 exhibits the morphological changes that occur during the hydration of MEA in cement slurry at different temperatures. The formation of Mg(OH)$_2$, which initiates at the surface of MEA particles and then progresses toward the center, is the primary product of this hydration process. An analysis of the figures indicates that the quantity of MEA hydration products increases with higher curing temperatures. For instance, in Figure 10a, the quantity of Mg(OH)$_2$ is limited, and its observation is challenging. Conversely, Figure 10c depicts a substantial amount of Mg(OH)$_2$ densely spread on the surface of spherical particles. Additionally, Figure 10e reveals the internal hydration of MEA particles, suggesting that MEA hydration is accelerated by higher temperatures. The figures also reveal that the density of the cement slurry increases with the curing temperature. Figure 10b clearly displays the pores generated during cement hydration, which are considerably reduced in Figure 10d,f.

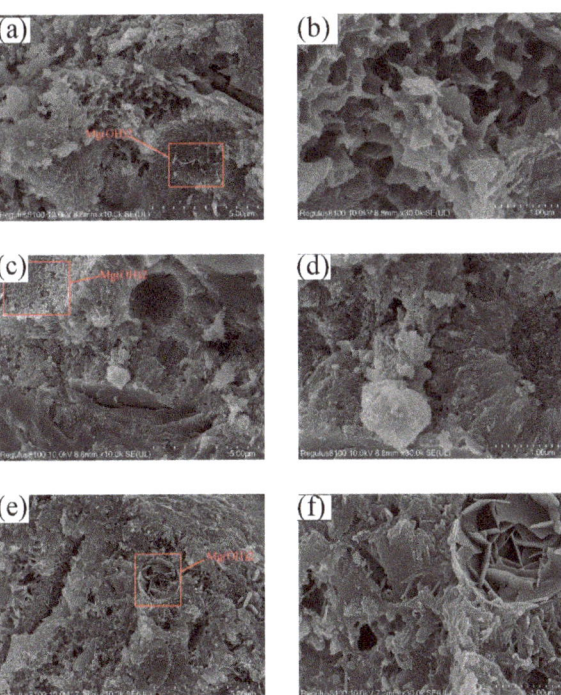

Figure 10. SEM images of hydration product in cement containing 8% MEA at 7 days (**a**) Microscopic morphology under temperature condition A; (**b**) Pore structure under temperature condition A; (**c**) Microscopic morphology under temperature condition B; (**d**) Microscopic morphology of Mg(OH)$_2$ under temperature condition B; (**e**) Microscopic morphology under temperature condition C; (**f**) Microscopic morphology of Mg(OH)$_2$ under temperature condition C.

3.5. Discussion

The impact of temperature on the performance of MEA is of great significance. The addition of MEA to concrete under temperature conditions B and C results in a larger expansion value compared to temperature condition A, which has a lower temperature peak. In temperature condition A, the early autogenous shrinkage of concrete cannot be fully compensated by the addition of 6% of MEA, while the addition of 8% of MEA only barely compensates for it. On the other hand, adding 6% of MEA at temperatures B and C leads to a significant expansion of the concrete. While the addition of 6%, 8%, and 10% of MEA does not differ much in expansion value at temperature A, there is a significant difference in expansion value at temperatures B and C with the addition of different amounts of MEA. Based on the thermogravimetry (TG) analysis, it is evident that the hydration of MEA is more complete at higher temperatures, which explains why the expansion of the concrete under temperature condition C is larger than under temperature conditions B and A. At temperature condition A, the hydration degree of MEA is only 30%, while at temperature condition B, the hydration degree of MEA has already reached over 60%. After 7 days of hydration at temperature condition C, the hydration degree of MEA is about 80%. This indicates that the hydration of MEA is rapid at temperatures above 60 °C, and the promoting effect is relatively weak before 60 °C, with the hydration rate not proportional to the temperature increase.

4. Conclusions

1. The effects of three different temperature conditions on the mechanical properties, autogenous shrinkage, and microstructure of high-strength concrete with MEA were studied. The main conclusions are as follows:
2. This investigation examined the impact of MEA incorporation on concrete's compressive strength under different temperature conditions. The results demonstrate that the addition of MEA to concrete at different levels results in a decrease in compressive strength, and this reduction becomes more significant with increased MEA content. Notably, the decline in compressive strength at a higher temperature C is less pronounced than that observed in temperatures A and B. This outcome is attributed to the rapid hydration of cement at high temperatures, which offsets the detrimental effects of MEA on the compressive strength of concrete. This suggests that adding a higher amount of MEA could be considered in engineering processes where higher temperatures are involved.
3. In comparison to temperature condition A, which exhibits a lower temperature peak value, the use of MEA in concrete mixtures under temperature conditions B and C resulted in significantly greater expansion values. Under temperature condition A, the addition of 6% MEA failed to fully offset early autogenous shrinkage in the concrete. When 6% of MEA was used in mixtures under temperature conditions B and C, significant expansion was observed. While mixtures containing 6%, 8%, and 10% of MEA exhibited minimal differences in expansion values under temperature condition A, there were substantial differences in expansion values observed under temperature conditions B and C for mixtures containing varying amounts of MEA. In previous construction projects, the addition of 8% MEA has been a common approach to mitigate the impact of shrinkage on concrete stability. However, this study suggests that a lower concentration of 6% MEA is sufficient to compensate for the shrinkage in two out of three tested temperature conditions. Therefore, it may be worthwhile to reduce the amount of MEA in order to maintain the stability and strength of concrete.
4. MEA hydration is more complete at elevated temperatures, which accounts for the greater expansion of concrete at temperature C compared to temperatures A and B. At temperature A, MEA achieves only 30% hydration, while at temperature B, it achieves more than 60%. The degree of hydration for MEA after 7 days at temperature C is approximately 80%. MEA undergoes rapid hydration at temperatures exceeding 60 °C, and the rate of hydration is not directly proportional to temperature. Further-

more, the enhancing effect of temperature on the hydration of MEA is relatively weak before 60 °C.
5. This paper investigated the compressive strength and autogenous deformation of MEA concrete under three temperature conditions. However, the actual temperature rise of mass concrete in engineering applications is diverse. More temperature conditions should be studied. In addition, this study only used one type of MEA, and more types of MEA can be investigated in the future.

Author Contributions: Conceptualization, M.D.; data curation, Z.C., X.H. and J.G.; writing—original draft preparation, Z.C.; writing—review and editing, M.D. and Z.M. All authors have read and agreed to the published version of the manuscript.

Funding: This work was supported by the Open Fund project of the National Laboratory for High Performance Civil Engineering Materials (2021CEM012) and the Priority Academic Program Development of Jiangsu Higher Education Institutions (PAPD).

Institutional Review Board Statement: Not applicable.

Informed Consent Statement: Not applicable.

Data Availability Statement: The data presented in this study are available on request from the corresponding author.

Acknowledgments: The authors gratefully acknowledge the assistance from Zhongyang Mao from NJTECH and the staff from the State Key Laboratory of Materials-Oriented Chemical Engineering.

Conflicts of Interest: The authors declare no conflict of interest.

References

1. Lu, X.; Chen, B.; Tian, B.; Li, Y.; Lv, C.; Xiong, B. A New Method for Hydraulic Mass Concrete Temperature Control: De-sign and Experiment. *Constr. Build. Mater.* **2021**, *302*, 124467. [CrossRef]
2. Zhou, H.; Zhou, Y.; Zhao, C.; Wang, F.; Liang, Z. Feedback Design of Temperature Control Measures for Concrete Dams Based On Real-Time Temperature Monitoring and Construction Process Simulation. *KSCE J. Civ. Eng.* **2018**, *22*, 1584–1592. [CrossRef]
3. Zhang, J. *Concrete Science*; Harbin Institute of Technology Press: Harbin, China, 2017.
4. Holt, E. Contribution of Mixture Design to Chemical and Autogenous Shrinkage of Concrete at Early Ages. *Cem. Concr. Res.* **2005**, *35*, 464–472. [CrossRef]
5. Li, Z.; Delsaute, B.; Lu, T.; Kostiuchenko, A.; Staquet, S.; Ye, G. A Comparative Study On the Mechanical Properties, Au-togenous Shrinkage and Cracking Proneness of Alkali-Activated Concrete and Ordinary Portland Cement Concrete. *Constr. Build. Mater.* **2021**, *292*, 1–11. [CrossRef]
6. Lee, K.M.; Lee, H.K.; Lee, S.H.; Kim, G.Y. Autogenous Shrinkage of Concrete Containing Granulated Blast-Furnace Slag. *Cem. Concr. Res.* **2006**, *36*, 1279–1285. [CrossRef]
7. Darquennes, A.; Staquet, S.; Delplancke-Ogletree, M.; Espion, B. Effect of Autogenous Deformation On the Cracking Risk of Slag Cement Concretes. *Cem. Concr. Compos.* **2011**, *33*, 368–379. [CrossRef]
8. Barcelo, L.; Moranville, M.; Clavaud, B. Autogenous Shrinkage of Concrete: A Balance Between Autogenous Swelling and Self-Desiccation. *Cem. Concr. Res.* **2005**, *35*, 177–183. [CrossRef]
9. EI, T. Chemical Shrinkage and Autogenous Shrinkage of Hydrating Cement Paste. *Cem. Concr. Res.* **1995**, *25*, 288–292.
10. Radocea, A. Autogenous Volume Change of Concrete at Very Early Age. *Mag. Concr. Res.* **1998**, *50*, 107–113. [CrossRef]
11. Lepage, S. Early Shrinkage Development in a High Performance Concrete. *Cem. Concr. Aggreg.* **1999**, *21*, 31–35.
12. Gao, P.; Lu, X.; Geng, F.; Li, X.; Hou, J.; Lin, H.; Shi, N. Production of MgO-type Expansive Agent in Dam Concrete by Use of Industrial By-Products. *Build. Environ.* **2008**, *43*, 453–457. [CrossRef]
13. Shen, D.; Liu, C.; Wen, C.; Kang, J.; Li, M.; Jiang, H. Restrained Cracking Failure Behavior of Concrete Containing MgO Compound Expansive Agent Under Adiabatic Condition at Early Age. *Cem. Concr. Compos.* **2023**, *135*, 104825. [CrossRef]
14. Mo, L.; Deng, M.; Tang, M.; Al-Tabbaa, A. MgO Expansive Cement and Concrete in China: Past, Present and Future. *Cem. Concr. Res.* **2014**, *57*, 1–12. [CrossRef]
15. Cao, F.; Miao, M.; Yan, P. Effects of Reactivity of MgO Expansive Agent On its Performance in Cement-Based Materials and an Improvement of the Evaluating Method of MEA Reactivity. *Constr. Build. Mater.* **2018**, *187*, 257–266. [CrossRef]
16. Mo, L.; Deng, M.; Tang, M. Effects of Calcination Condition On Expansion Property of MgO-type Expansive Agent Used in Cement-Based Materials. *Cem. Concr. Res.* **2010**, *40*, 437–446. [CrossRef]
17. Chen, X.; Yang, H.; Li, W. Factors Analysis On Autogenous Volume Deformation of MgO Concrete and Early Thermal Cracking Evaluation. *Constr. Build. Mater.* **2016**, *118*, 276–285. [CrossRef]

18. Ferrini, V.; De Vito, C.; Mignardi, S. Synthesis of Nesquehonite by Reaction of Gaseous CO_2 with Mg Chloride Solution: Its Potential Role in the Sequestration of Carbon Dioxide. *J. Hazard. Mater.* **2009**, *168*, 832–837. [CrossRef]
19. Liu, J.P.; Wang, Y.J.; Tian, Q.; Zhang, S.Z. Modeling Hydration Process of Magnesia Based On Nucleation and Growth Theory: The Isothermal Calorimetry Study. *Thermochim. Acta* **2012**, *550*, 27–32. [CrossRef]
20. Li, H.; Tian, Q.; Zhao, H.; Lu, A.; Liu, J. Temperature Sensitivity of MgO Expansive Agent and its Application in Temper-ature Crack Mitigation in Shiplock Mass Concrete. *Constr. Build. Mater.* **2018**, *170*, 613–618. [CrossRef]
21. Sant, G. The Influence of Temperature On Autogenous Volume Changes in Cementitious Materials Containing Shrinkage Reducing Admixtures. *Cem. Concr. Compos.* **2012**, *34*, 855–865. [CrossRef]
22. Al, T.Q.T.Y. Temperature Sensitivity Analysis On Expansion Property of MgO Composite Expansion Agent. *Water Power* **2010**, *36*, 49–51.
23. Liu Jiaping, W.Y. Temperature Sensitivity of Light Calcined Magnesia Expansion Agent and its Mechanism Analysis. *J. Southeast Univ. (Nat. Sci. Ed.)* **2011**, *41*, 359–364.
24. Yang Dongyang, C.H. Effect of MgO Expansive Agent on Shrinkage Performance of Ultra-High Performance Concrete. *Bull. Chin. Ceram. Soc.* **2022**, *41*, 3420–3427.
25. *GB/T 50081-2019*; Standard for Test Methods of Concrete Physical and Mechanical Properties. Standards Press of China: Beijing, China, 2019.
26. *GB/T 22314-2008*; Plastics—Epoxide Resins—Determination of Viscosity. Standards Press of China: Beijing, China, 2008.
27. Gao, P.; Wu, S.; Lu, X.; Deng, M.; Lin, P.; Wu, Z.; Tang, M. Soundness Evaluation of Concrete with MgO. *Constr. Build. Mater.* **2007**, *21*, 132–138. [CrossRef]
28. Li, Y.; Deng, M.; Mo, L.; Tang, M. Strength and Expansive Stresses of Concrete with MgO-type Expansive Agent Under Restrain Conditions. *J. Build. Mater.* **2012**, *15*, 446–450.
29. Wang, L.; Li, G.; Li, X.; Guo, F.; Tang, S.; Lu, X.; Hanif, A. Influence of Reactivity and Dosage of MgO Expansive Agent On Shrinkage and Crack Resistance of Face Slab Concrete. *Cem. Concr. Compos.* **2022**, *126*, 104333. [CrossRef]
30. Zhang, J. Recent Advance of MgO Expansive Agent in Cement and Concrete. *J. Build. Eng.* **2022**, *45*, 103633. [CrossRef]
31. Yin, C.; Zong, W.; Wang, S.; Li, G.; Li, Q.; Lu, L. Effect of MgO On Composition, Structure and Properties of Alite-Calcium Strontium Sulphoalminate Cement. *J. Chin. Ceram. Soc.* **2011**, *39*, 20–24.
32. Yao, F.; Deng, M.; Mo, L. Expanion of Three- Graded Concrete Containing MgO Expansive Agent Cured by 80 °C Steam. *Concrete* **2012**, *08*, 34–40.
33. Mo, L.; Fang, J.; Huang, B.; Wang, A.; Deng, M. Combined Effects of Biochar and MgO Expansive Additive On the Au-togenous Shrinkage, Internal Relative Humidity and Compressive Strength of Cement Pastes. *Constr. Build. Mater.* **2019**, *229*, 116877. [CrossRef]
34. Li, S.; Mo, L.; Deng, M.; Cheng, S. Mitigation On the Autogenous Shrinkage of Ultra-High Performance Concrete Via Using MgO Expansive Agent. *Constr. Build. Mater.* **2021**, *312*, 125422. [CrossRef]

Disclaimer/Publisher's Note: The statements, opinions and data contained in all publications are solely those of the individual author(s) and contributor(s) and not of MDPI and/or the editor(s). MDPI and/or the editor(s) disclaim responsibility for any injury to people or property resulting from any ideas, methods, instructions or products referred to in the content.

Article

Effect of the Water-Binder Ratio on the Autogenous Shrinkage of C50 Mass Concrete Mixed with MgO Expansion Agent

Jun Chen [1], Zhongyang Mao [1], Xiaojun Huang [1] and Min Deng [1,2,*]

[1] College of Materials Science and Engineering, Nanjing Tech University, Nanjing 211816, China; 202061103064@njtech.edu.cn (J.C.)

[2] State Key Laboratory of Material-Oriented Chemical Engineering, Nanjing Tech University, Nanjing 211800, China

* Correspondence: dengmin@njtech.edu.cn; Tel.: +86-136-0518-4865

Abstract: The high adiabatic temperature rise and low heat dissipation rate of mass concrete will promote rapid hydration of the cementitious material and rapid consumption of water from the concrete pores, which may significantly accelerate the development of concrete autogenous shrinkage. In this study, the effect of the water-binder ratio on the autogenous shrinkage of C50 concrete mixed with MgO expansion agent (MEA) was explained with respect to mechanical properties, pore structure, degree of hydration, and micromorphology of the concrete based on a variable temperature curing chamber. The results show that the high temperature rise within the mass concrete accelerates the development of early (14 d) autogenous shrinkage of the concrete, and that the smaller the water-binder ratio, the greater the autogenous shrinkage of the concrete. With the addition of 8 wt% MEA, the autogenous shrinkage of concrete can be effectively compensated. The larger the water-binder ratio, the higher the degree of MgO hydration, and in terms of the compensation effect of autogenous shrinkage, the best performance is achieved at a water-binder ratio of 0.36. This study provides a data reference for the determination of the water-binder ratio in similar projects with MEA.

Keywords: MgO expansion agent; autogenous shrinkage; C50 mass concrete

1. Introduction

High-strength concrete is being used more widely due to the increasing demands of modern construction [1,2]. Compared with ordinary concrete, high-strength concrete has a lower water-binder ratio (typically ≤ 0.4) due to the addition of high-efficiency water-reducing agents [3–5]. When the water-binder ratio of concrete is less than 0.40 [6], autogenous shrinkage accounts for a greater proportion of the total shrinkage, which is an important factor causing concrete cracking [7–10]. The study of autogenous shrinkage in high-strength concrete needs more attention as it shortens the service life of concrete structures.

A great deal of research has been carried out and solutions have been proposed, both nationally and internationally, on how to reduce the autogenous shrinkage of concrete [11–13]. There are currently two mainstream methods. One method is to add lightweight aggregates [14–18], or highly absorbent polymers [19,20] and other internal curing materials to the concrete for internal curing, in order to reduce the rate of self-drying inside the concrete and thus reduce the autogenous shrinkage of high-strength concrete. However, there are few practical engineering applications due to the disadvantages of reducing the strength of concrete and the high cost. The second method is to add an expansion agent to the concrete to compensate for the autogenous shrinkage of the concrete through the expansion effect produced by the expansion agent. There are three main types of concrete expansion agents commonly used in practical engineering: sulfate of alumina, calcium oxide, and magnesium oxide [21,22]. The sulfate of alumina expansive agent requires a large amount of water—in concrete with a low water-binder ratio, its expansion effect cannot be fully achieved. Moreover, its hydration product, calcium alumina, is easy to

decompose at high temperatures, and there is a risk of expansion collapsing. Calcium oxide expansion agents' hydration is too fast, and the effect of shrinkage compensation for aged concrete is weak. On the contrary, magnesium oxide expansion agents have the advantages of low water requirements, stable products, and controlled expansion rates; therefore, they are often used to compensate for the long-term shrinkage of mass concrete [23].

Under normal temperature conditions, the use of MEA to compensate for autogenous shrinkage of concrete has been well documented [24,25]. However, in some large volume straight wall concrete with special applications, the poor thermal conductivity of bulk concrete and the large amount of heat of hydration given off by the hydration of cement can cause the temperature of the structural concrete to rise rapidly, which may significantly accelerate the development of concrete autogenous shrinkage [26–28] and affect the use of MEA. Therefore, the effect of temperature must be considered when using MEA to compensate for shrinkage. The effect of maintenance temperature on the expansion of MEA on concrete has been extensively studied [29–32]. However, MgO has a strong temperature sensitivity [33] and the expansion effect of MEA is influenced by the internal temperature history of bulk concrete, so the performance of MEA cannot be predicted according to a constant temperature. Li et al. [34] investigated the synergistic effect of MEA and internal curing materials on the shrinkage compensation of concrete under variable temperature conditions, relying on urban railway projects. Wang et al. [35] studied the compensating effect of a magnesium oxide expander with different activity and different dosing levels on the shrinkage of panel concrete. Li et al. [36] demonstrated the expansion performance of a magnesium oxide expander with different reactivity in ship lock concrete and combined it with different temperature histories. Ou et al. [22] used a self-developed temperature stress testing machine to study the crack mitigation properties of different reactive magnesium oxide expansion agents in concrete under different engineering conditions. These studies show that the compensatory effect of MEA on shrinkage varies in different projects. The current application of MEA in engineering is mostly focused on its reactivity and the amount of admixture, ignoring the effect of concrete mixing water on the compensatory shrinkage performance of MEA at variable temperatures. At room temperature, the autogenous shrinkage is higher with the reduction of the water-binder ratio [37,38]. Cao et al. [39] proved that the larger the water-binder ratio, the better the compensation effect when using MEA to compensate for concrete shrinkage at room temperature. However, based on the temperature history of actual projects, there are still few reports on whether the effect of temperature on the autogenous shrinkage of different water-binder ratios—or the effect on the expansion of MEA—can have a cross-matched effect, and what kind of water-binder ratio can achieve the best effect. This may lead to the incorrect use of MEA in practical engineering.

This study is based on a large volume straight wall concrete project, using the same concrete mix ratio as that of the straight wall on site (8 wt% of MEA). Through a simulation of the site temperature history (high temperature peak and low cooling rate), the effect of the water-binder ratio on the autogenous shrinkage of concrete with MEA under the variable working temperature conditions was investigated, in order to provide a basis for the selection of the water-binder ratio of concrete for engineering structures.

2. Materials and Methods

2.1. Materials

In this research, Ordinary Portland cement (OPC) was provided by Jiangnan Onoda Cement Co., Ltd. (Nanjing, Jiangsu Province, China). Fly Ash (FA) was supplied from Nanjing, and blast furnace slag (BFS) was provided by Mabel Ltd. (Nanjing, Jiangsu Province, China), which were used as the supplementary cementitious materials. The MEA was supplied by Nanjing Subote New Material Co., Ltd. (Nanjing, Jiangsu Province, China), of which the reactivity value was 200 s. Its reactivity was estimated by the citric acid method in accordance with Chinese standard CBMF-2017 [40]. The chemical compositions of OPC, FA, BSF, and MEA are provided in Table 1. Distributions of particle size of these cementitious materials are presented in Figure 1. A polycarboxylate superplasticizer (SP), provided by Nanjing Subote New Material Co., Ltd., was used for maintaining a consistent slump of concrete, whose water reduction rate was 30% and solid content was 40%.

Table 1. Chemical compositions of raw materials.

Chemical Compositions	OPC	FA	BFS	MEA
CaO (%)	65.32	4.07	38.00	1.98
SiO_2 (%)	18.55	50.53	33.72	3.87
Al_2O_3 (%)	3.95	31.65	17.74	1.03
Fe_2O_3 (%)	3.41	4.48	0.77	0.88
MgO (%)	1.01	0.92	6.35	89.37
K_2O (%)	0.72	1.26	0.41	0.08
Na_2O (%)	0.18	0.68	0.41	-
SO_3 (%)	2.78	1.32	1.04	0.06
LOI (%)	2.88	2.77	−0.72	2.38

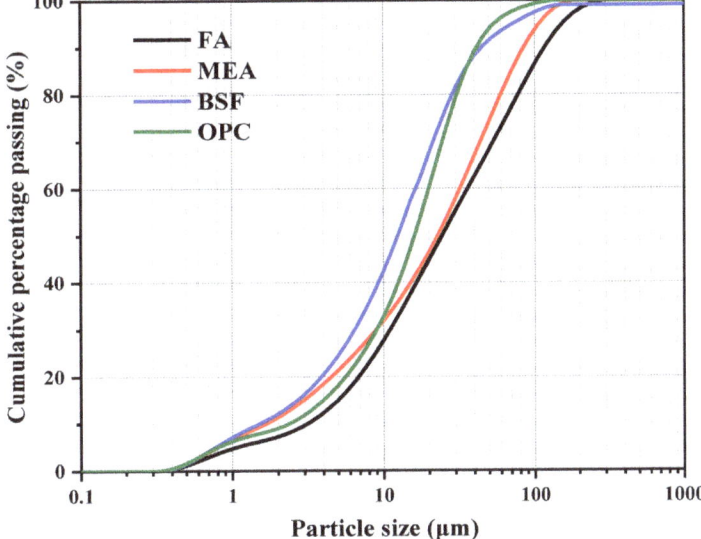

Figure 1. Particle size distribution of raw materials.

2.2. Mixture Proportions of Concrete

Four concrete mixtures (numbered from C1 to C4) containing 8 wt% MEA are shown in Table 2, whose water-binder ratios range from 0.28 to 0.40. Mixtures C5–C8 were designed as references, which were mixed without MEA in this research. In mixtures C1–C4, FA, BFS, and MEA were used as a partial replacement for cement in order to reduce the total hydration heat release and the temperature peak of bulk concrete, which may decrease the high risk of concrete cracking. The amount of sand (S) and gravel (G) used is also shown in Table 2, and the sand percentage is 38%. The quantity of MEA is determined according to the actual project, in which the addition of MEA at 8 wt% of the total cementitious materials restrained the cracking of concrete walls, whereas 6 wt% did not. This is also supported by the research of Yu [41].

Table 2. Mixture ratio of concrete (kg/m³).

Sample	W/B	Mix Proportion/(kg/m³)						
		OPC	FA	BFS	S	G	MEA	SP
C1	0.28	360	90	40	680	1100	40	3.3%
C2	0.32	360	90	40	680	1100	40	2.6%
C3	0.36	360	90	40	680	1100	40	1.9%
C4	0.40	360	90	40	680	1100	40	1.6%
C5	0.28	320	90	40	680	1100	0	3.2%
C6	0.32	320	90	40	680	1100	0	2.5%
C7	0.36	320	90	40	680	1100	0	1.7%
C8	0.40	320	90	40	680	1100	0	1.4%

2.3. Methods

2.3.1. Mechanical Properties

A variable temperature curing chamber was used to simulate the history of changes in the central temperature of straight walls in engineering. Figure 2a presents a typical temperature history profile for large volume straight wall concrete of civil engineering, which has a high temperature rise and low cooling rate. The central temperature of the concrete achieves a temperature peak at around 30 h and drops to room temperature at around 14 days. Figure 2b presents the curve of temperature change simulated in the laboratory.

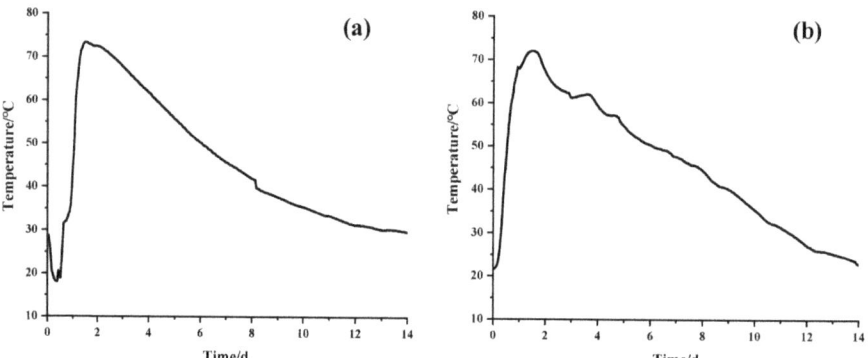

Figure 2. Temperature curves: (a) temperature history of straight walls; (b) temperature history of the test chamber simulation.

According to Chinese standard GB/T 50081-2019 [42], the concrete was poured into molds whose dimensions were 150 mm × 150 mm × 150 mm. Cured in the variable temperature oven for 24 h, concrete specimens would be removed from molds and wrapped in plastic wrap, after which they continued to be cured. The compressive strength of concrete was tested at 3 d, 7 d, and 28 d—and an average for each sample, at each age, was taken from triplicate tests.

2.3.2. Autogenous Deformation

Considering that the concrete of the straight walls in the project was in a sealed state before the early demolding, which would have exacerbated the autogenous drying shrinkage, the autogenous deformation test was performed. The test for autogenous shrinkage of concrete is carried out using the buried strain gauge method with reference to SL/L 352-2020 [43]. After the concrete mixing was completed, fresh concrete was cast into a cylindrical PVC pipe (Ø 160 mm × 400 mm) and a W-15 sine-type strain gauge, as well as a moisture meter, was buried in the middle part of the specimen (Figure 3). Finally, the concrete surface was sealed with paraffin wax. The concrete mixtures loaded in PVC pipes were cured by the variable temperature curing chamber according to the above temperature curve, and the shrinkage was recorded every half hour over 28 days. The final setting time of the concrete was set as the "zero time" for autogenous shrinkage [44].

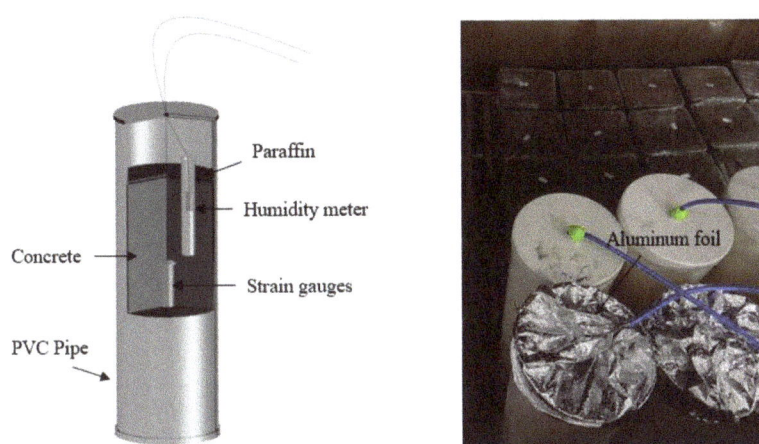

Figure 3. PVC pipe and PVC pipe in a variable temperature curing chamber.

2.3.3. Thermal Analysis

At 1 d, 3 d, and 7 d, the mixtures of concrete without sand were analyzed to calculate the hydration of magnesium oxide. After being cut into pieces, the samples were soaked in the anhydrous ethanol for a week for the purpose of terminating their hydration. Then, the samples were placed in a vacuum drying oven at 60 °C until the mass was stable. After, those dried samples were ground into powders of no more than 0.080 μm. The powder samples were tested using NETZSCH STA 449F1 thermogravimetry under a N_2 atmosphere, heated from room temperature to 950 °C.

2.3.4. Pore Structure

The pore structure of concrete specimens was tested using a mercury intrusion porosimeter (MIP). According to the sample preparation requirements, the samples maintained in the variable temperature oven were taken out. Concrete specimens were cut into 3~5 mm pieces and soaked in anhydrous ethanol, which was aimed at terminating the hydration of samples. Afterward, the samples were placed in an oven at the temperature of 60 °C until the mass was stable.

3. Results

3.1. Effect of Water-Binder Ratio on the Compressive Strength of Concrete with MEA

Figure 4 illustrates the compressive strength of the concrete with different water-binder ratios at different ages. As indicated in Figure 5a, the compressive strength under variable temperature conditions of curing was consistently over 50 MPa at 28 d, and more than 85% of the 28 d strength could be achieved at 7 d. Similar to normal temperature curing, the compressive strength of concrete decreases as the water-binder ratio increases. Compared with the concrete without MEA, which is presented in Figure 4b, MEA hardly affects the compressive strength of concrete. Before 7 d, the compressive strength of the concrete with MEA is slightly lower than that of the concrete without MEA. However, when the concrete is cured for 28 days, both are almost equal in compressive strength. This may be attributed to MEA replacing part of the cement, which lessens the contribution of the hydration products of MgO to strength [45]. In addition, MgO of MEA will compete with cement for water in the early stage, affecting the early hydration of cement; hence, the early strength of concrete mixed with MEA is lower than that of concrete not mixed with MEA. As the reaction proceeds, the slurry structure becomes denser, and the compressive strength of MEA-doped concrete does not differ significantly from that of non-MEA-doped concrete.

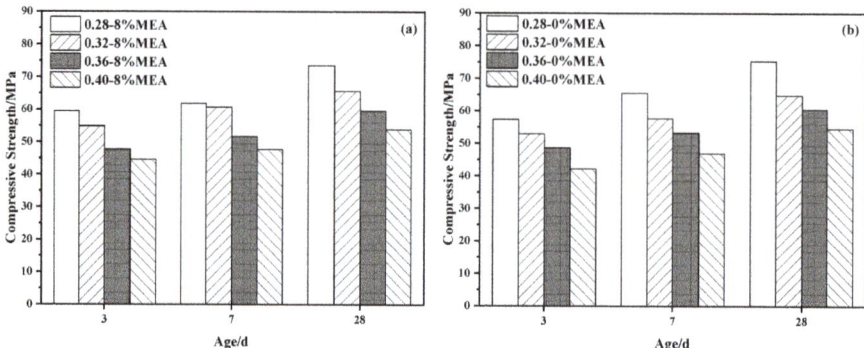

Figure 4. Compressive strength of the concrete with different water-binder ratios: (**a**) with MEA; (**b**) without MEA.

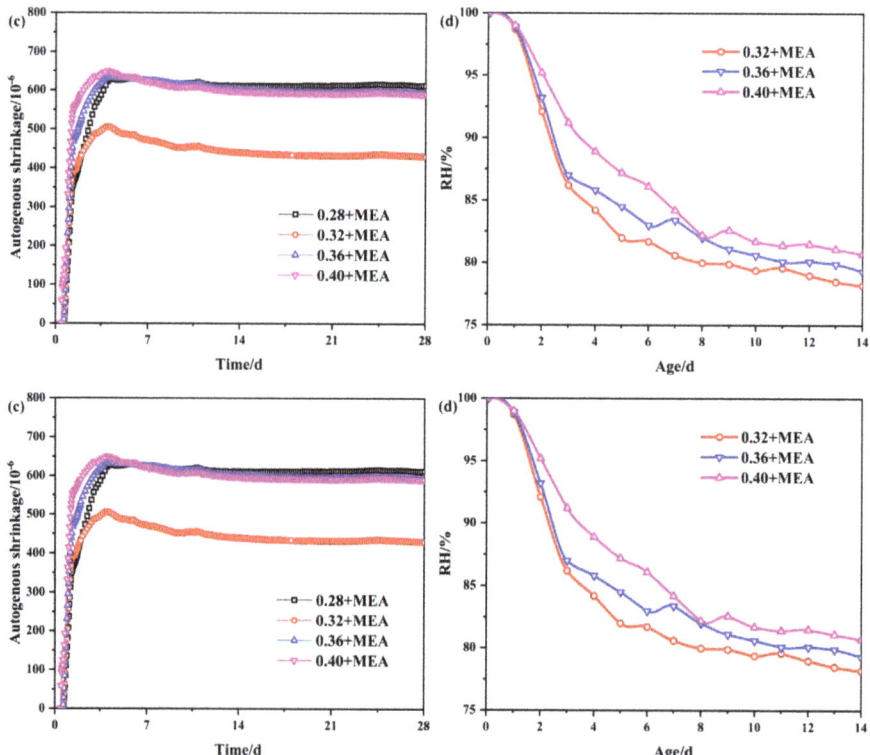

Figure 5. Effect of water-binder ratios on the autogenous deformation and internal relative humidity of concrete: (**a**) autogenous deformation, without MEA; (**b**) internal relative humidity, without MEA; (**c**) autogenous deformation, with MEA; (**d**) internal relative humidity, with MEA.

3.2. Autogenous Deformation and Internal Relative Humidity in Concrete

Figure 5 represents the autogenous deformation of concrete with different water-binder ratios under variable temperatures. As demonstrated in Figure 5a, the autogenous deformation of concrete without MEA shows a tendency toward a smaller water-binder ratio, the larger the autogenous deformation. Concrete whose water-binder ratio is 0.28 had the greatest autogenous deformation at 18 d, up to −218 microstrain. However, when the water-binder ratio was 0.40, the concrete produced 38 microstrain expansion. This confirms the direct link between autogenous deformation of concrete and internal moisture, and it can be concluded that the water-binder ratio of 0.40 is the cut-off at which the issue of autogenous deformation should be taken into account. After 14 d of curing, the autogenous shrinkage tends to steady. The results of the compressive strength of the concrete show that there is little difference between the 7 d compressive strength and the 28 d strength of the concrete. This means that most of the hydration has been completed at 7 d under the simulated temperature history. The self-drying caused by hydration is the main reason for the autogenous shrinkage of the concrete, and therefore, the autogenous shrinkage of the concrete tends to be stable after 14 d. Figure 5c shows the autogenous deformation of concrete with different water-binder ratios after the addition of 8 wt% MEA. As shown in Figure 5c, the autogenous deformation of concrete consistently shows expansion, especially before 7 d, in the period when the autogenous deformation of concrete develops rapidly. After 7 d, it reaches a stable stage of development, and within 28 d, it produces an inverse shrinkage within 100 microstrain. The autogenous deformation of the concrete samples with the water-binder ratios of 0.36 and 0.40 are essentially the same at 28 d, at around 600 microstrain. Compared to the concrete without MEA, that with MEA has a better compensation effect on autogenous shrinkage when the water-binder ratio is 0.36.

Figure 5b,d show the internal humidity curves of concrete for different water-binder ratios. As presented in the subfigures, the addition of MEA has little effect on the internal humidity of concrete. However, it is worth noting that the internal humidity of concrete decreases significantly during 1–7 d, which corresponds to the rapid development of early autogenous shrinkage of concrete. After being cured for 14 d, the humidity slowly decreases, which also explains why the autogenous shrinkage of concrete remains near stable after 14 d. The variation in the internal humidity of the concrete essentially matches the variation of the curing temperature, with high temperatures accelerating the depletion of moisture within the concrete and the development of autogenous deformation.

3.3. Thermal Analysis

Figure 6 shows the TG and DSC profiles of the concrete specimens with different water-binder ratios. As shown in Figure 6b,d,f, there is an obvious endothermic peak that appears at 320–400 °C, which corresponds to the decomposition of $Mg(OH)_2$. Based on Mo [46], the quantity of $Mg(OH)_2$ and the hydration degree of MgO can be calculated by Equations (1) and (2):

$$Q_{Mg(OH)_2} = \frac{58 \times \text{Mass loss}_{(320\ °C - 400\ °C)}}{18} \quad (1)$$

$$H_{MgO} = \frac{40 \times \text{Mass loss}_{(320\ °C - 400\ °C)}}{n \times 18 \times \left[1 - \text{Mass loss}_{(950\ °C)}\right]} \quad (2)$$

Table 3 provides the amount of $Mg(OH)_2$ and the hydration degree of MgO in the samples. It is clear that there is a rapid increase in the hydration degree of MgO during the first 3 d. The water-binder ratios (0.28, 0.32, 0.36, 0.40) are 53.1%, 52.3%, 55.2%, and 67.4%, respectively, which are highly dependent on the curing temperature at which the specimens are maintained, once again confirming the temperature sensitivity of the MEA and that high temperatures accelerate the hydration of MgO. The hydration degrees of MgO in concrete with different water-binder ratios are relatively close at 3 d, which is probably attributable to the fact that there is sufficient water inside the concrete in the early stages and MgO competes with cement for water. When at 7 d, there is a notable difference between different water-binder ratios. The hydration degrees of MgO in the concrete with the water-binder ratios of 0.40, 0.36, and 0.40 are 72.6%, 66.9%, and 62.8%, respectively, whereas the hydration degree is only 58.9% when the water-binder ratio is 0.28. Combined with the changes of humidity in Figure 5, the reduction in moisture in the concrete impedes the hydration of MgO. The lower the water-binder ratio, the less free water is available for MgO, or the lower the rate. It is obvious that internal moisture becomes critical in the later stages of curing.

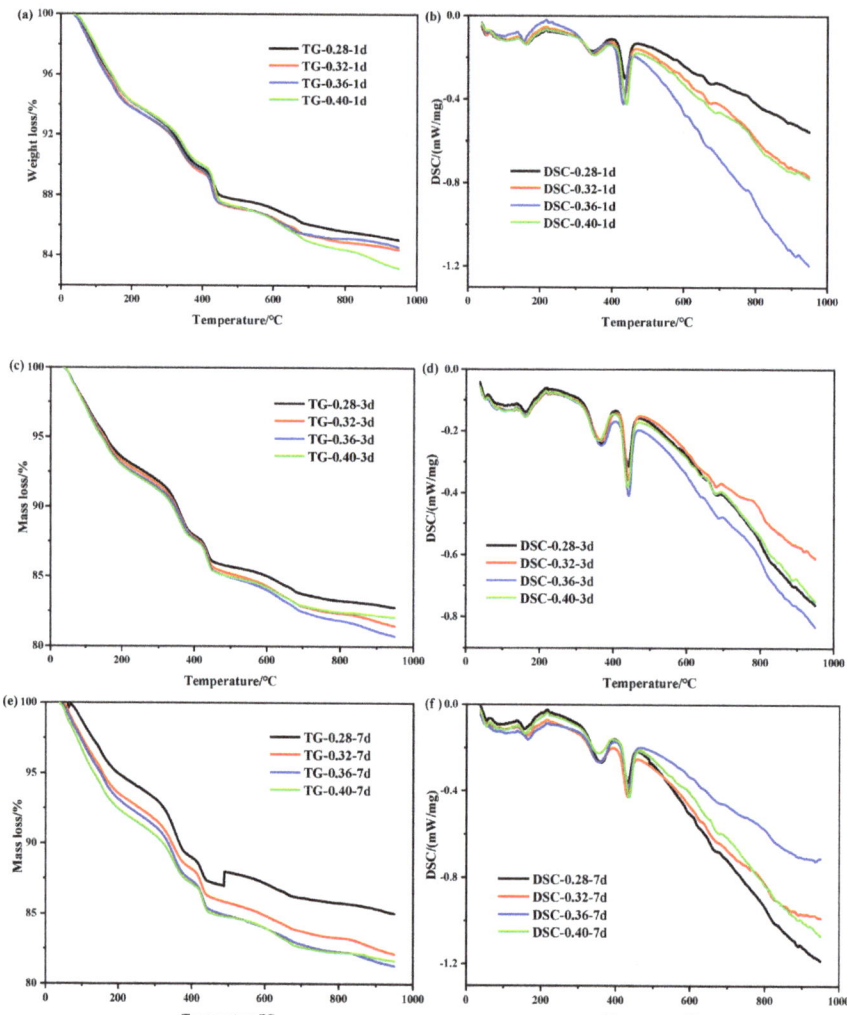

Figure 6. TG/DSC curves of the concrete containing MEA with different water-binder ratios: (**a**) TG curves at the ages of 1 d; (**b**) DSC curves at the ages of 1 d; (**c**) TG curves at the ages of 3 d; (**d**) DSC curves at the ages of 3 d; (**e**) TG curves at the ages of 7 d; (**f**) DSC curves at the ages of 7 d.

Table 3. Quantity of Mg(OH)$_2$ and hydration degree of MgO in concrete pastes containing MEA.

	0.28 + 8% MEA			0.32 + 8% MEA			0.36 + 8% MEA			0.40 + 8% MEA		
	1 d	3 d	7 d	1 d	3 d	7 d	1 d	3 d	7 d	1 d	3 d	7 d
Mass loss at 320–400 °C/wt%	2.1	2.8	3.1	2.2	2.9	3.2	2.2	3.2	3.3	2.2	3.6	3.4
Quantity of Mg(OH)$_2$/%	6.8	9.3	10.0	7.2	9.5	10.4	8.2	10.5	10.7	7.2	11.7	11.1
Hydration degree of MgO/%	39.2	53.1	58.9	40.5	52.3	65.8	42.7	55.2	66.9	51.7	67.4	72.6

3.4. Pore Structure and Microstructure

Figure 7 shows the pore structures of concrete specimens mixed with MEA (8 wt%) and different water-binder ratios at 3 d. As illustrated in Figure 7a, the pore size of the samples added with MEA is mostly distributed in the range of 7~70 nm. As the water-binder ratio increases, the peak of the pore size moves to the right and the pore size < 50 nm increases, probably due to the higher hydration degree of the silicate cement and the formation of more gel pores. Figure 7b indicates that the cumulative porosity of the specimens is also increased. Figure 8 shows the relative pore volume distribution of MgO concrete with various water-binder ratios. It is obvious that the larger the water-binder ratio, the fewer pores there are above >50 nm. The increase in water contributes to the migration of Mg$^+$, which allows more Mg(OH)$_2$ products to grow inside the pores, thus changing the pore size distribution inside the pastes and refining the pore size.

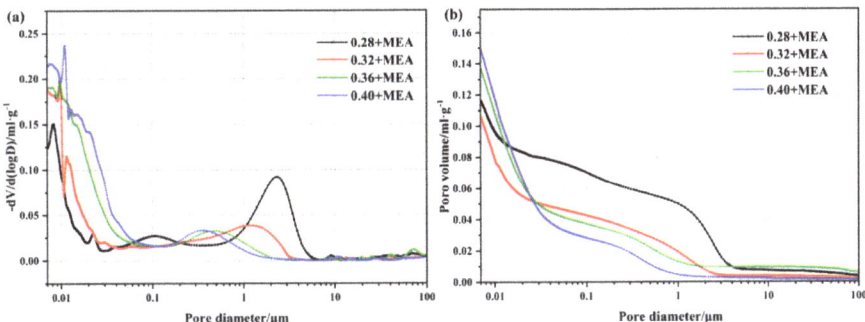

Figure 7. Pore size distributions of the concrete containing MEA with different water–binder ratios at 3 d: (a) incremental volume; (b) cumulative volume.

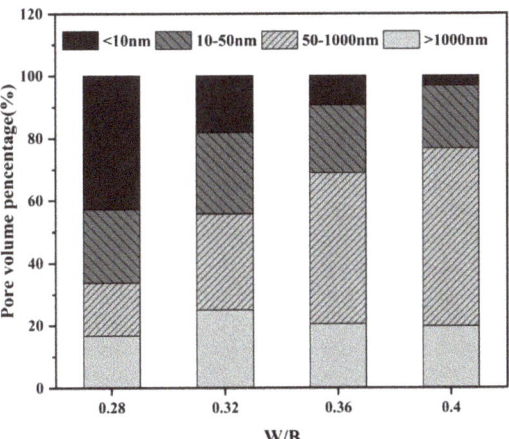

Figure 8. Relative pore volume distribution of concrete pastes.

Figure 9 presents the microscopic morphology of the specimens with the water-binder ratios of 0.32 and 0.36. As illustrated in Figure 9b, under high temperature maintenance, the Mg(OH)$_2$ crystals are mainly in the form of flakes of different sizes, which grow near the MgO particles. Ca^{2+} has an inhibitory effect on the diffusion of Mg^{2+} and promotes the growth of Mg(OH)$_2$; therefore, Mg(OH)$_2$ crystals generally grow near Ca(OH)$_2$ crystals. The lamellar Mg (OH)2 crystals in Figure 9d are denser and fill in the pores, which explains the smaller pore size of the specimen with a water-binder ratio of 0.36 compared to that with a ratio of 0.28.

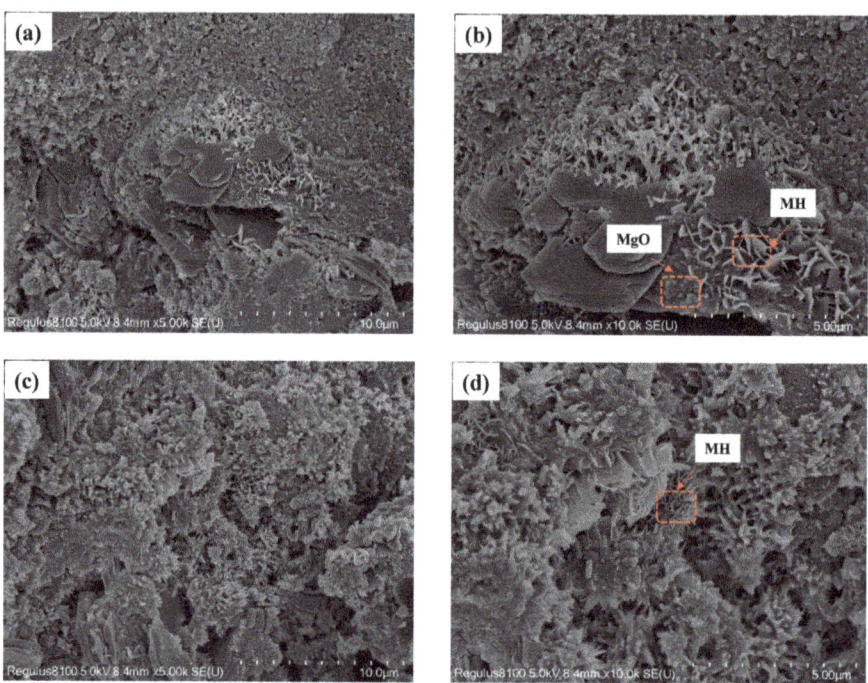

Figure 9. SEM images of hydrated MgO in concrete pastes at 7 days: (**a**) w/c = 0.32, ×5000; (**b**) w/c = 0.32, ×10,000; (**c**) w/c = 0.36, ×5000; (**d**) w/c = 0.36, ×10,000.

4. Discussion

Bulk high-strength concrete generally has a high adiabatic temperature rise due to the great amount of cementitious material and the slow dissipation of heat from the concrete. In large volume straight wall concrete projects, temperatures in the center of the concrete can reach over 70 °C (and even 80 °C) and do not drop to room temperature until 14 d. The high temperature changes within the bulk straight wall concrete not only accelerate the development of autogenous concrete deformation but also promote the hydration of MgO. Although the addition of MEA mitigates the autogenous shrinkage of the concrete, the effect of different water-binder ratios on the deformation of concrete under variable temperature conditions varies slightly. Compared to concrete with a water-binder ratio of 0.28, an increased water-binder ratio not only reduces the autogenous shrinkage of the concrete itself but also contributes more to the hydration of MgO and the production of more Mg (OH)$_2$. The thermal analysis shows that there is little difference in the degree of hydration of MgO in concrete with different water-binder ratios at 3 d. However, at 7 d, the hydration degree of MgO in concrete with a water-binder ratio of 0.40 is 72.6% compared to 58.9% with a water-binder ratio of 0.28. Sufficient water is supplied to MgO to produce Mg (OH)$_2$, which counteracts the pore pressure and is the key to compensating the autogenous shrinkage of concrete. Meanwhile, Ca^{2+} has an inhibitory effect on the diffusion of Mg^{2+}, which promotes the nucleation and growth of Mg (OH)$_2$. Thus, Mg (OH)$_2$ crystals mostly grow overlappingly in pores not far from the MgO particles. Moreover, the higher the water-binder ratio, the denser the hydration product of MgO. The early arrival of the hydration period of MgO allows the crystalline growth pressure of the Mg (OH)$_2$ crystals to resist the early shrinkage deformation of the concrete. However, the autogenous deformation of mass concrete in practical projects is more complex, as the restraint of reinforcement, the large temperature difference between day and night, and the method of construction all affect the autogenous shrinkage of concrete. Increasing or changing the water-binder ratio alone will not completely overcome the problem of autogenous deformation of concrete and needs to be carried out in cooperation with other effective measures according to the actual situation of the project at the time. This study was carried out based on a large volume of straight wall concrete and is intended to provide a reference for similar projects with MEA.

5. Conclusions

The internal temperature variations of C50 bulk straight wall concrete structures in practical projects were simulated in the laboratory. On this basis, the influence of different water-binder ratios on the compressive strength, autogenous shrinkage, pore size, and micromorphology of concrete with MEA was studied. The key conclusions are the following:

1. The hydration process of MgO is accelerated under variable temperature curing. The addition of 8 wt% MEA reduces the compressive strength of the concrete, but the effect is minimal (no more than 5 MPa). In this experiment, the compressive strength of the concrete at 7 d can reach 85% or more at 28 d.
2. High temperature accelerates the early self-shrinkage of concrete, especially within 7 d. Increasing the water-binder ratio can reduce the autogenous shrinkage of concrete. When the water-binder ratio is 0.40, the concrete deformation is slightly expanded. Compared to the reference group without MEA, the best effect of compensating for autogenous shrinkage is achieved by adding 8% MEA with a water-binder ratio of 0.36. Therefore, in actual projects, the water-binder ratio can be relaxed to 0.36 or even 0.40 under the premise of ensuring strength.
3. At high temperatures, the hydration product of MgO, Mg(OH)$_2$ crystals, appear as lamellar in the cement system, growing mainly in the pores not far from the MgO particles and overlapping each other. As the water-binder ratio increases, the MgO hydrates more and the Mg(OH)$_2$ crystals become denser, which fills the pores better and counteracts the autogenous shrinkage of the concrete.

Author Contributions: Conceptualization, M.D.; data curation, J.C. and X.H.; writing—original draft preparation, J.C.; writing—review and editing, M.D. and Z.M. All authors have read and agreed to the published version of the manuscript.

Funding: This work was supported by the Open Fund project of the National Laboratory for High-Performance Civil Engineering Materials (2021CEM012) and the Priority Academic Program Development of Jiangsu Higher Education Institutions (PAPD).

Institutional Review Board Statement: Not applicable.

Informed Consent Statement: Not applicable.

Data Availability Statement: The data presented in this study are available on request from the corresponding author.

Acknowledgments: The authors gratefully acknowledge the assistance from Zhongyang Mao and Xiaojun Huang from N/TECH, and the staff from the State Key Laboratory of Materials-Oriented Chemical Engineering.

Conflicts of Interest: The authors declare no conflict of interest.

References

1. Habel, K.; Viviani, M.; Denarié, E.; Brühwiler, E. Development of the mechanical properties of an Ultra-High-Performance Fiber Reinforced Concrete (UHPFRC). *Cem. Concr. Res.* **2006**, *36*, 1362–1370. [CrossRef]
2. Sifan, M.; Nagaratnam, B.; Thamboo, J.; Poologanathan, K.; Corradi, M. Development and prospectives of lightweight high strength concrete using lightweight aggregates. *Constr. Build. Mater.* **2023**, *362*, 129628. [CrossRef]
3. Mac, M.J.; Yio, M.H.N.; Wong, H.S.; Buenfeld, N.R. Analysis of autogenous shrinkage-induced microcracks in concrete from 3D images. *Cem. Concr. Res.* **2021**, *144*, 106416. [CrossRef]
4. Jensen, O.M.; Hansen, P.F. Autogenous deformation and RH-change in perspective. *Cem. Concr. Res.* **2001**, *31*, 1859–1865. [CrossRef]
5. Yio, M.H.N.; Mac, M.J.; Yeow, Y.X.; Wong, H.S.; Buenfeld, N.R. Effect of autogenous shrinkage on microcracking and mass transport properties of concrete containing supplementary cementitious materials. *Cem. Concr. Res.* **2021**, *150*, 106611. [CrossRef]
6. Tang, S.; Huang, D.; He, Z. A review of autogenous shrinkage models of concrete. *J. Build. Eng.* **2021**, *44*, 103412. [CrossRef]
7. Zhao, H.; Liu, J.; Yin, X.; Wang, Y.; Huang, D. A multiscale prediction model and simulation for autogenous shrinkage deformation of early-age cementitious materials. *Constr. Build Mater.* **2019**, *215*, 482–493. [CrossRef]
8. Shen, D.; Liu, C.; Luo, Y.; Shao, H.; Zhou, X.; Bai, S. Early-age autogenous shrinkage, tensile creep, and restrained cracking behavior of ultra-high-performance concrete incorporating polypropylene fibers. *Cem. Concr. Compos.* **2023**, *138*, 104948. [CrossRef]
9. Valipour, M.; Khayat, K.H. Coupled effect of shrinkage-mitigating admixtures and saturated lightweight sand on shrinkage of UHPC for overlay applications. *Constr. Build Mater.* **2018**, *184*, 320–329. [CrossRef]
10. Wang, Y.; Tian, Q.; Li, H.; Wang, Y.; Li, M.; Liu, J. A new hypothesis for early age expansion of cement-based materials: Cavitation in ink-bottle pores. *Constr. Build. Mater.* **2022**, *326*, 126884. [CrossRef]
11. Akcay, B.; Tasdemir, M.A. Autogenous shrinkage, pozzolanic activity and mechanical properties of metakaolin blended cementitious materials. *KSCE J. Civil Eng.* **2019**, *23*, 4727–4734. [CrossRef]
12. Wang, L.; Jin, M.; Wu, Y.; Zhou, Y.; Tang, S. Hydration, shrinkage, pore structure and fractal dimension of silica fume modified low heat Portland cement-based materials. *Construct. Build. Mater.* **2021**, *272*, 121952. [CrossRef]
13. Weber, S.; Reinhardt, H.W. A New Generation of High Performance Concrete: Concrete with Autogenous Curing. *Adv. Cem. Based Mater.* **1997**, *6*, 59–68. [CrossRef]
14. Andreu Gonzalez-Corominas, M.E. Effects of using recycled concrete aggregates on the shrinkage of high performance concrete. *Constr. Build. Mater.* **2016**, *115*, 32–41. [CrossRef]
15. Torgal, F.P.; Jalali, S. *Eco-Efficient Construction and Building Materials*; Springer Verlag London Limited: London, UK, 2011.
16. Polat, R.; Demirboğa, R.; Khushefati, W.H. Effects of nano and micro size of CaO and MgO, nano-clay and expanded perlite aggregate on the autogenous shrinkage of mortar. *Constr. Build Mater.* **2015**, *81*, 268–275. [CrossRef]
17. Yang, L.; Ma, X.; Liu, J.; Hu, X.; Wu, Z.; Shi, C. Improving the effectiveness of internal curing through engineering the pore structure of lightweight aggregates. *Cem. Concr. Comp.* **2022**, *134*, 104773. [CrossRef]
18. Xu, F.; Lin, X.; Zhou, A. Performance of internal curing materials in high-performance concrete: A review. *Constr. Build Mater.* **2021**, *311*, 125250. [CrossRef]
19. Shen, D.; Wang, X.; Cheng, D.; Zhang, J.; Jiang, G. Effect of internal curing with super absorbent polymers on autogenous shrinkage of concrete at early age. *Constr. Build Mater.* **2016**, *106*, 512–522. [CrossRef]
20. Wyrzykowski, M.; Lura, P.; Pesavento, F.; Gawin, D. Modeling of Water Migration during Internal Curing with Superabsorbent Polymers. *J. Mater. Civ. Eng.* **2012**, *24*, 1006–1016. [CrossRef]
21. Mo, L.; Deng, M.; Tang, M.; Al-Tabbaa, A. MgO expansive cement and concrete in China: Past, present and future. *Cem. Concr. Res.* **2014**, *57*, 1–12. [CrossRef]
22. Ou, G.; Lin, Z.; Kishi, T. The practical application of a self-developed temperature stress testing machine in development of expansive cement blended with calcium sulfoaluminate additives. *Cem. Concr. Res.* **2023**, *164*, 107045. [CrossRef]
23. Ou, G.; Kishi, T.; Mo, L.; Lin, Z. New insights into restrained stress and deformation mechanisms of concretes blended with calcium sulfoaluminate and MgO expansive additives using multi-scale techniques. *Constr. Build. Mater.* **2023**, *371*, 130737. [CrossRef]
24. Chen, X.; Yang, H.; Li, W. Factors analysis on autogenous volume deformation of MgO concrete and early thermal cracking evaluation. *Constr. Build Mater.* **2016**, *118*, 276–285. [CrossRef]
25. Temiz, H.; Kantarcı, F.; Emin İnceer, M. Influence of blast-furnace slag on behaviour of dolomite used as a raw material of MgO-type expansive agent. *Constr. Build Mater.* **2015**, *94*, 528–535. [CrossRef]
26. Zhang, X.; Liu, Z.; Wang, F. Autogenous shrinkage behavior of ultra-high performance concrete. *Constr. Build Mater.* **2019**, *226*, 459–468. [CrossRef]
27. Loukili, A.; Khelidj, A.; Richard, P. Hydration kinetics, change of relative humidity, and autogenous shrinkage of ultra-high-strength concrete. *Cem. Concr. Res.* **1999**, *29*, 577–584. [CrossRef]
28. Persson, B. Self-desiccation and its importance in concrete technology. *Mater. Struct.* **1997**, *30*, 293–305. [CrossRef]
29. Salomão, R.; Bittencourt, L.R.M.; Pandolfelli, V.C. A novel approach for magnesia hydration assessment in refractory castables. *Ceram Int.* **2007**, *33*, 803–810. [CrossRef]

30. Mo, L.; Fang, J.; Hou, W.; Ji, X.; Yang, J.; Fan, T.; Wang, H. Synergetic effects of curing temperature and hydration reactivity of MgO expansive agents on their hydration and expansion behaviours in cement pastes. *Constr. Build Mater.* **2019**, *207*, 206–217. [CrossRef]
31. Durán, T.; Pena, P.; De Aza, S.; Gómez-Millán, J.; Alvarez, M.; De Aza, A.H. Interactions in Calcium Aluminate Cement (CAC)-Based Castables Containing Magnesia. Part I: Hydration-Dehydration Behavior of MgO in the Absence of CAC. *J. Am. Ceram Soc.* **2011**, *94*, 902–908. [CrossRef]
32. Thomas, J.J.; Musso, S.; Prestini, I. Kinetics and Activation Energy of Magnesium Oxide Hydration. *J. Am. Ceram Soc.* **2014**, *97*, 275–282. [CrossRef]
33. Liu, J.; Wang, Y.; Tian, Q.; Zhang, S. Temperature sensitivity of the expansion performance of lightly burned magnesium oxide expansion agent and its mechanism analysis. *J. Southeast Univ.* **2011**, *41*, 359–364.
34. Li, M.; Liu, J.; Tian, Q.; Wang, Y.; Xu, W. Efficacy of internal curing combined with expansive agent in mitigating shrinkage deformation of concrete under variable temperature condition. *Constr. Build. Mater.* **2017**, *145*, 354–360. [CrossRef]
35. Wang, L.; Li, G.; Li, X.; Guo, F.; Tang, S.; Lu, X.; Hanif, A. Influence of reactivity and dosage of MgO expansive agent on shrinkage and crack resistance of face slab concrete. *Cem. Concr. Compos.* **2022**, *126*, 104333. [CrossRef]
36. Li, H.; Tian, Q.; Zhao, H.; Lu, A.; Liu, J. Temperature sensitivity of MgO expansive agent and its application in temperature crack mitigation in shiplock mass concrete. *Constr. Build. Mater.* **2018**, *170*, 613–618. [CrossRef]
37. Miao, M.; Yan, P. Influences of Water-binder Ratio and Dosage of Fly Ash on Autogenous Shrinkage of Shrinkage-Compensating Concrete. *J. Chin. Ceram. Soc.* **2012**, *40*, 1607–1612.
38. Yang, L.; Shi, C.; Wu, Z. Mitigation techniques for autogenous shrinkage of ultra-high-performance concrete—A review. *Compos. Part B Eng.* **2019**, *178*, 107456. [CrossRef]
39. Cao, F.; Miao, M.; Yan, P. Effects of reactivity of MgO expansive agent on its performance in cement-based materials and an improvement of the evaluating method of MEA reactivity. *Constr. Build. Mater.* **2018**, *187*, 257–266. [CrossRef]
40. *CBMF 19-2017*; China Building Materials Association Standard. MgO Expansion Agent for Concrete. 2017. Available online: http://www.ccpa.com.cn/site/xhbz/93.html?siteid=10001 (accessed on 28 February 2022).
41. *GB/T 50081-2019*; Standard for Test Methods of Concrete Physical and Mechanical Properties. China Construction Industry Press: Beijing, China, 2019.
42. Yu, L.; Deng, M.; Mo, L.; Liu, J.; Jiang, F. Effects of Lightly Burnt MgO Expansive Agent on the Deformation and Microstructure of Reinforced Concrete Wall. *Adv. Mater Sci. Eng.* **2019**, *2019*, 1–9. [CrossRef]
43. *SL/T 352-2020*; Test Code for Hydraulic Concrete. China Water & Power Press: Beijing, China, 2020.
44. Huang, H.; Ye, G. Examining the "time-zero" of autogenous shrinkage in high/ultra-high performance cement pastes. *Cem. Concr. Res.* **2017**, *97*, 107–114. [CrossRef]
45. Li, Y.; Deng, M.; Mo, L.; Tang, M. Strength and Expansive Stresses of Concrete with MgO-Type Expansive Agent under Restrain Conditions. *J. Build. Mater.* **2012**, *15*, 446–450.
46. Mo, L.; Fang, J.; Huang, B.; Wang, A.; Deng, M. Combined effects of biochar and MgO expansive additive on the autogenous shrinkage, internal relative humidity and compressive strength of cement pastes. *Constr. Build. Mater.* **2019**, *229*, 116877. [CrossRef]

Disclaimer/Publisher's Note: The statements, opinions and data contained in all publications are solely those of the individual author(s) and contributor(s) and not of MDPI and/or the editor(s). MDPI and/or the editor(s) disclaim responsibility for any injury to people or property resulting from any ideas, methods, instructions or products referred to in the content.

Article

Preparation and Performance of Repair Materials for Surface Defects in Pavement Concrete

Pengfei Li [1], Zhongyang Mao [1,2], Xiaojun Huang [1,2] and Min Deng [1,2,*]

[1] College of Materials Science and Engineering, Nanjing Tech University, Nanjing 211800, China; 202061203171@njtech.edu.cn (P.L.)
[2] State Key Laboratory of Materials-Oriented Chemical Engineering, Nanjing 211800, China
* Correspondence: dengmin@njtech.edu.cn; Tel.: +86-136-0518-4865

Abstract: Concrete surface defects are very complex and diverse, which is a great test for repair materials. The efficiency and durability of the repair system depend on the bonding effect between the concrete and the repair material. However, the rapid increase in system viscosity during the reaction of repair materials is an important factor affecting the infiltration effect. In the present work, the infiltration consolidation repair material was prepared, and its basic properties (viscosity, surface drying time and actual drying time, infiltration property) and mechanical properties were evaluated. Finally, the infiltration depth, film-forming thickness, and anti-spalling ability of concrete under a single-side freeze–thaw cycle are revealed. The results showed that using ethyl acetate could rapidly reduce the viscosity of the repair material, and the repair material could penetrate 20–30 mm into the concrete within 10 min. It was found by laser confocal microscopy that the thickness of the film formation after 3 days was only 29 μm. In the mortar fracture repair test to evaluate the bond strength, the bond strength of the repaired material reached 9.18 MPa in 28 days, and the new fracture surface was in the mortar itself. In addition, the freeze–thaw cycle test was carried out on the composite specimens under salt solution to verify the compatibility of the designed repair material with the concrete substrate. The data showed that the average amount of spalling was only 1704.4 g/m^2 when 10% ethyl acetate was added. The penetrating repair material in this study has good infiltration performance, which can penetrate a certain depth in the surface pores and form a high-performance consolidation body, forming a "rooted type" filling.

Keywords: surface defect; repair material; ethyl acetate; infiltration performance

1. Introduction

With the dramatic increase in traffic volume in recent years, concrete pavement bears more and more load. The long-term load and impact pressure, coupled with the destruction of the natural environment (freeze–thaw cycles, dry–wet cycles, corrosive substances), concrete pavement withstands varying degrees of damage [1]. Once the defects appear, they provide a channel for the corrosive medium, aggravate the failure of the concrete, and seriously reduce the service quality of the pavement [2–4]. Excavating and re-paving the damaged road surface inevitably causes a waste of manpower and material resources as well as an inconvenience to open traffic and fast navigation [5,6]. Therefore, it is an urgent problem to repair concrete pavement quickly and efficiently. Selecting appropriate repair materials can effectively delay the sustainable development of concrete micro-cracks and improve the durability of concrete, which requires the repair materials to have low viscosity, excellent mechanical properties, and strength stability [7–9].

At present, the materials used for concrete repair are mainly divided into three categories. The first category is inorganic repair materials which mainly include silica cement, fast-hardening Portland cement, magnesium phosphate cement, and so on [10,11]. The second category is organic repair materials which mainly include epoxy resin, acrylic resin, polyurethane, and acrylamide [12]. The third category is polymer repair materials compounded by organic and inorganic materials, for example, vinyl polyester acid emulsion modified cement mortar, propyl cement mortar, and epoxy mortar concrete [13,14]. However, there are many kinds of repair materials, and they all have different advantages and disadvantages.

The research on concrete repair materials now focuses on good compatibility with concrete, good durability, and low viscosity for easy grouting and repair. Kim et al. [15] used a combination of epoxy resin injection and surface pretreatment by a dipping agent. For the freeze–thaw cycle test, both the relative dynamic elastic modulus and durability index were greater than 80% after 300 cycles when the exposure ended for all the specimens, and the chloride resistance test results confirmed that repair methods, including impregnation, had a penetration effect 2–3.5 times greater than the epoxy repair method. Hassan et al. [16] used an acrylic copolymer and a cement/aggregate/admixture mixture to form a polymer-modified repair mortar (PMC). The PMC repair mortar showed the most appropriate properties in terms of dimensional stability with concrete due to a similar elastic modulus and low shrinkage strains when compared to the parent concrete. Khalina et al. [17] studied the effect of aliphatic reactive diluents on the properties of two different epoxy resins, bifunctional and multifunctional epoxy resins. It was found that as the diluent was gradually incorporated into the epoxy resin, the damage mode of the epoxy resin specimens changed from brittle to more ductile under the bending test. The fracture strain of the bifunctional resin increased by 75%, and that of the multifunctional resin increased by more than 24%. The diluent content enhanced the fracture toughness of the epoxy resin by about 20% and 29% for the bifunctional and multifunctional resins, respectively.

As surface defects contain many tiny cracks, it is difficult for high-viscosity organic repair materials to penetrate the surface layer to a certain depth. Moreover, the higher the viscosity, the poorer the wettability of repair materials, most of which remain on the concrete surface to form thick polymers [18–21]. At present, the viscosity of organic repair materials is high, and it is difficult to penetrate a certain depth of concrete surface, and it is unable to complete the effective plugging of pores [22]. In addition, the surface cures to form a thick polymer film, which is easily removed, resulting in a poor repair effect. Therefore, reducing the viscosity of the prepolymer is important to ensure excellent permeability.

In view of this, tetrachloroethylene was selected as a shrinkage reducer [23], and anhydrous ethanol, ethyl acetate, and acetone were used as inactive diluents. The influence of diluent addition on the material properties (viscosity, grouting depth, gel time) was clarified, and a series of studies were completed on the mechanical properties of the materials. This paves the way for penetrating repair materials to repair pavement defects.

2. Materials and Methods

2.1. Raw Materials

The design strength grade of an ordinary concrete specimen is C30, in which the cement is Conch brand P·O 42.5 ordinary Portland cement. The aggregate is a well-graded crushed stone with a particle size of less than 30 mm and a fineness modulus of 2.5 medium sand. See Table 1 for the specific mix ratio.

Table 1. Proportioning of C30 Concrete.

Water	Cement	Sand	Coarse Aggregate
205 kg/m^3	340 kg/m^3	705 kg/m^3	1150 kg/m^3

Methyl methacrylate (MMA) is the main raw material, dibenzoyl peroxide (BPO) is the initiator, dibutyl phthalate (DBP) is the plasticizer, tetrachloroethylene (PCE) is the shrinkage agent, N,N-dimethylaniline (DMA) is the curing agent, and the three diluents are absolute ethanol, ethyl acetate, and acetone. Table 2 shows the proportion of raw materials used to prepare MMA-based repair materials.

Table 2. Proportion of raw materials.

NO.	MMA /g	BPO /%	DBP /%	DMA /%	PCE /%	Anhydrous Ethanol /%	Ethyl Acetate /%	Acetone /%
KB	80	1.5	30	0.5	10	0	0	0
W5	80	1.5	30	0.5	10	5	0	0
W10	80	1.5	30	0.5	10	10	0	0
W15	80	1.5	30	0.5	10	15	0	0
E5	80	1.5	30	0.5	10	0	5	0
E10	80	1.5	30	0.5	10	0	10	0
E15	80	1.5	30	0.5	10	0	15	0
A5	80	1.5	30	0.5	10	0	0	5
A10	80	1.5	30	0.5	10	0	0	10
A15	80	1.5	30	0.5	10	0	0	15

2.2. Preparation of the Penetrating Repair Materials

1. Preparation of the MMA-based repair materials

MMA is used as the main raw material, BPO as the catalyst, DBP as the plasticizer, and PCE as the shrinkage agent to modify the repair material. The preparation process is as follows.

- A certain amount of MMA was poured into a beaker, stirred at a speed of 300/min, and distilled in a water bath at 50 °C for 10 min.
- After distillation to remove the inhibitor, 1.5% BPO, 30% DBP, and 10% PCE was added and continuously stirred for 2 min.
- The curing agent DMA was added to the prepolymer and stirred evenly until the set time of the test was reached. The three-flask was quickly put into cold water to slow down the polymerization reaction, and the MMA-based repair material was obtained.

2. Preparation of the MMA-based penetrating repair materials

Infiltration-type repair materials penetrate a certain depth of concrete surface by self-weight. In order to ensure good infiltration performance of the repair materials, thinner is used to reduce the viscosity of the repair materials, and DMA is used to promote curing to prepare the penetrating repair materials. The preparation process is as follows.

- The configured prepolymer was added with a certain amount of DMA and stirred for 2 min at 300 r/min.
- After an average of 60 min, diluent was added and stirred at 300 r/min for 3 min to obtain permeable repair material.

2.3. Pretreatment of the Concrete Specimen

A number of 100 × 100 × 100 mm concrete specimens were prepared and placed in a laboratory with a temperature of (20 ± 2) °C and a relative humidity of (65 ± 5)% for 28 days. The cured specimens adopt the single-side freeze–thaw method according to the long-term performance and durability test method of ordinary concrete (GBT50082-2009) so that the infiltration surface of the repaired material is closer to the damaged concrete with surface defects. The specific specimen is shown in Figure 1.

Figure 1. Surface of concrete specimen after pretreatment.

2.4. Test Methods

2.4.1. Viscosity Test

The initial viscosity of the repair material has a great influence on the infiltration performance. The smaller the viscosity of the repair material, the stronger its fluidity. In this study, the NDJ-1 rotary viscometer (Shenzhen, China) was used to measure the viscosity of the repair material. The viscometer equipment is shown in Figure 2. By analyzing the ratio of raw materials in Table 2, the initial viscosity of the repair material was measured at 20 °C ± 2 °C, and the average viscosity of 60 min was used as the evaluation index.

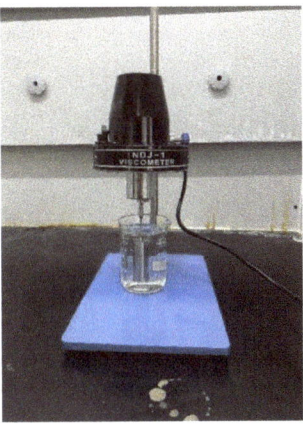

Figure 2. Pointer rotary viscometer NDJ-1.

2.4.2. Surface-Drying Time and Through-Drying Time

The surface-drying time is determined according to the provisions of GB/T 1728-2020 method B (refers to trigger), and the surface of the coating film is considered dry when no paint adheres to the finger after touching the surface of the coating film. The through-drying time is determined according to the provisions of GB/T 1728-2020 in method C (blade method), and the coating film is considered dry by cutting and scraping the coating film on the template with a blade and observing that there is no adhesion phenomenon in the bottom layer and film.

2.4.3. Determination of the Film-Forming Thickness

Film forming thickness is an important index to evaluate the surface properties of the tunnel. The thinner the film thickness is, the less influence it has on the original excellent friction property of the concrete surface. In this study, a laser confocal microscope (OLYMPUS, Japan) was used to observe a section of the concrete specimen after 6 h, 24 h, and 3 d of curing of the coated repair material at 25 °C. The equipment is shown in Figure 3. The curing specimens were cut in the middle, and the average thickness of five uniform points in the section was taken as the film thickness.

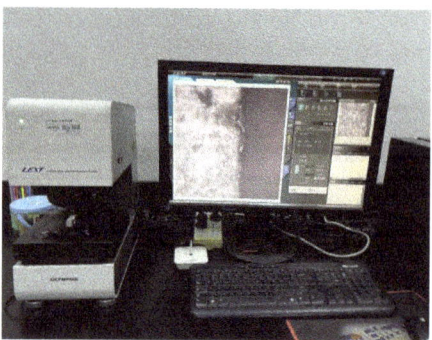

Figure 3. Laser confocal microscope.

2.4.4. Infiltration Performance

The infiltration performance of the repair material is evaluated by measuring the infiltration depth. A 100 × 100 × 100 mm concrete specimen is used, and the concrete surface is first cleaned of floating ash before applying the repair material. The repair material is applied three times with a 10-min interval between each application. The repaired specimen is then cured at room temperature for 3 d. The concrete block is cut, and the infiltration of the solidified repair material is observed from the cut surface. Three points with relatively uniform infiltration are selected, and the infiltration depth at each point is observed. The average of the infiltration depths at the three points is taken as the final infiltration depth of the repair material on the surface of the concrete.

2.4.5. Bond Strength

Using P·O 42.5 cement, mortar specimens with dimensions of 40 × 40 × 160 mm were prepared. The specimens were cured under standard conditions (temperature of (20 ± 2) °C and relative humidity of 95%) for 28 d. After curing, the specimens were placed in a drying oven at 105 °C for 6 h to serve as the dry interface specimens for bond strength testing. The specimens that had completed standard curing were used as the wet interface specimens for bond strength testing. The mortar flexural strength testing machine was used to break the specimens, and according to the Japanese standard JISA6024-1998, a gap of 3~5 mm was left between the two broken sections of the specimen and fixed in place. MMA prepolymer was then injected into the gap [24]. As shown in Figure 4, after being cured in a 60 °C oven for 4 h, the repaired material experienced some volume shrinkage. Therefore, the cracks were filled again with MMA prepolymer. The flexural strength of the specimens was tested, and the fracture location was observed after being cured at room temperature (temperature of (20 ± 2) °C and relative humidity of 60%) for 3 d, 7 d, and 28 d.

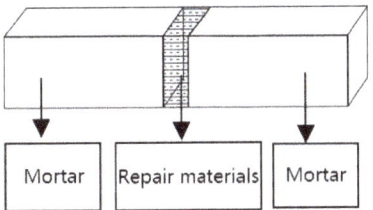

Figure 4. The model view of specimen for bond test.

2.4.6. Freeze–Thaw Cycle Test

The freeze–thaw cycle test after concrete repair can reflect the performance of the repaired material in an environment of large temperature difference [25]. According to the long-term performance and durability of ordinary concrete test method GBT50082-2009, this test adopts a single-side freeze–thaw method. The specimens that have reached the specified curing age are placed in a laboratory with a temperature of (20 ± 2) °C and a relative humidity of (65 ± 5)% to dry for 28 d. The side of the 100 × 100 × 100 mm specimen is coated with Sikaflex neutral silicone sealant and sealed with aluminum foil paper. All sides are sealed to prevent side peeling and ensure that the specimens are in a single-sided water absorption state. When adding the test liquid (3% NaCl solution) to the test container, care was taken to avoid wetting the top of the specimen, which is in a capillary absorption state. The specimens are immersed in NaCl solution to a depth of about 10 mm. Each group of tests consists of 5 specimens, and measurements are taken every four cycles. Ultrasonic cleaning is used to collect the peeling materials, as shown in Figure 5a,b for specific operating steps.

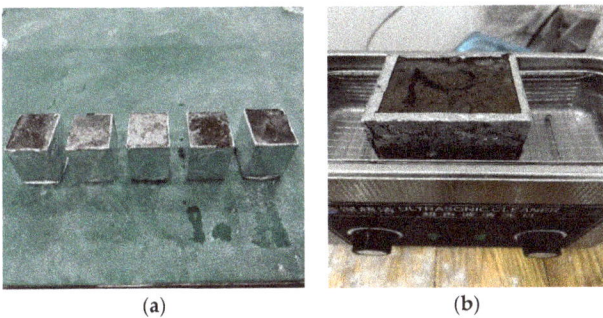

Figure 5. Preparation of freeze–thaw specimens: (**a**) prepared specimen; (**b**) ultrasonic cleaning.

The calculation and treatment of test results comply with the following provisions:

$$\mu_s = \mu_b - \mu_f \tag{1}$$

The mass of crushed concrete debris on the surface of the specimen μ_s is calculated according to equation (1), where μ_s is the quality of concrete slag on the specimen surface; μ_f is the quality of filter paper; and μ_b is the total mass of filter paper and concrete slag after drying.

After N freeze–thaw cycles, the total mass of detached particles per unit testing surface area of an individual specimen is calculated using Equation (2).

$$m_n = \frac{\sum \mu_s}{A} \times 10^6 \tag{2}$$

where m_n is the N freeze–thaw cycles, single specimen unit test surface area total mass of exfoliation (g/m^2); μ_s is the spalling mass of the specimen obtained from each test gap; and A is the surface area of a single specimen test surface (mm^2).

The arithmetic mean value of the total spalling mass per unit area of each group of specimens was taken as the measurement value of the total spalling mass m_n.

2.4.7. SEM

A JSM-IT200 scanning electron microscope (Beijing, China) was used to observe the high-resolution microstructure of cement mortar. The equipment is shown in Figure 6. The samples used were taken from the broken parts of the mortar surface and scanned point by point with a focused high-energy electron beam, magnifying 10–100,000 times.

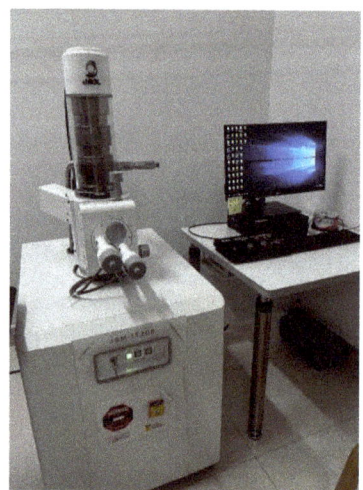

Figure 6. Scanning electron microscope.

3. Results and Discussion

3.1. Preparation Technology of the MMA-Based Repair Materials

In this study, a curing agent was mainly added to accelerate the curing of repair materials. Therefore, the content of the curing agent has a great impact on the properties of the repair materials. In order to obtain the best preparation technology for infiltration consolidation repair materials, the average viscosity at 60 min, surface drying time, and tensile strength were used as evaluation indexes. The experiment investigated the addition of 0.3%, 0.5%, 0.7%, and 1% of the curing agent to the prepolymer with the proportion of MMA 60, BPO 1.5%, DBP 30%, and PCE 10%.

3.1.1. Effect of Preparation Technology on the 60-Min Average Viscosity

The initial viscosity of the repair material is an important factor affecting the repair effect. Using the average viscosity after 60 min of adding the curing agent as the evaluation index, Figure 7 shows the variation range of the average viscosity of the repair material after adding different amounts of curing agent.

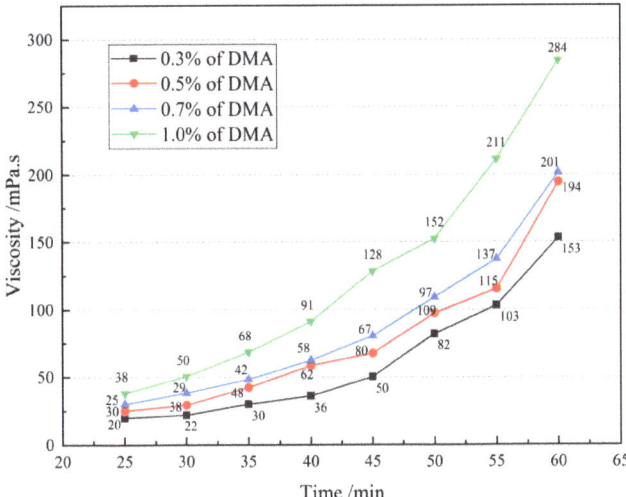

Figure 7. 60-min average viscosity range.

As shown in Figure 7, the 60-min average viscosity of the repair material generally showed a slow-then-fast trend. In the early stage of polymerization, when the synthesis time is short, the number of activated centers for MMA monomers is small, and the polymerization chains are not long, resulting in a small polymerization rate. Therefore, the viscosity of the repair material does not change much with increasing synthesis time. At a reaction time of 25 min, the viscosity of the repair material with 0.3% DMA content is 20 mPa·s, while that of the repair material with 1% DMA content is 38 mPa·s. Within the time interval of 25 to 40 min, the viscosity curve of the material shows a steady increase, maintaining below 100 mPa·s. As the reaction progresses from 40 to 60 min, the viscosity of the system increases significantly. The viscosity of the repair material with 1% DMA content increases sharply, and the average viscosity at 60 min reaches 284 mPa·s, which is a 46.12% increase compared to the average viscosity of 153 mPa·s for the repair material with 0.3% DMA content. Therefore, DMA content has a significant impact on the viscosity of the repair material. When the concentration of the curing agent is too high, the polymerization reaction rate will accelerate, generating a large amount of heat. If it cannot dissipate in time, the viscosity will suddenly increase, causing explosive polymerization, seriously affecting the repair effect.

3.1.2. Effect of Preparation Technology on the 60-Min Drying Time

The drying time of the repair material directly affects the work performance and can also greatly shorten the construction period. The prepolymer is applied to the template and cured at 25 °C. The drying time and actual drying time of the repair material were discussed, and the effect of different curing agent dosages on the drying time and actual drying time of the material was analyzed, as shown in Figure 8.

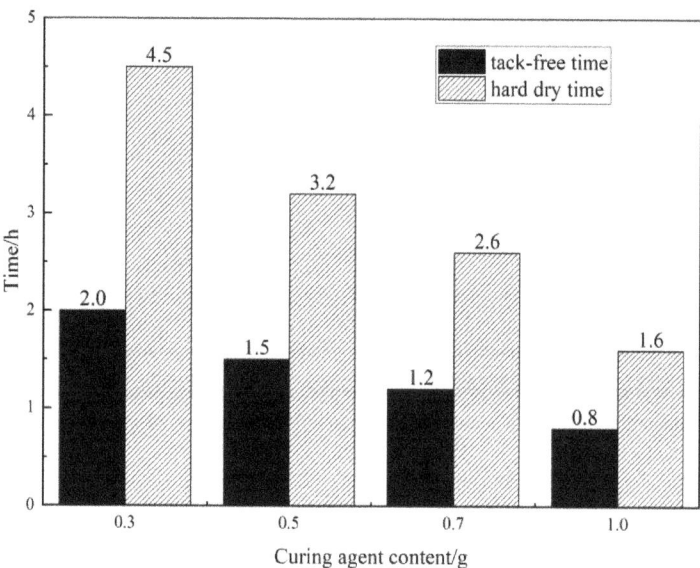

Figure 8. Influence of curing agent content on drying time.

From Figure 8, it can be observed that the drying time and actual drying time of the material decrease with an increase in the curing agent content. When the DMA content is 1%, the drying time and actual drying time of the repair material reach 0.8 h and 1.5 h, respectively, which are 60 and 67% shorter than the drying time and actual drying time of the repair material with 0.3% DMA content. It can be seen that the proportion of the curing agent in the MMA-based repair materials is the main factor affecting the curing rate of the system. When the amount of curing agent added is too small, it cannot participate in the reaction fully, or the performance of the cured product is poor and cannot meet the requirements. On the other hand, when the amount of curing agent added is too large, the curing rate of the system increases significantly, resulting in the production of a large number of bubbles during curing, which seriously affects the physical properties of the material itself. Therefore, in practical applications, the amount of curing agent should be adjusted reasonably according to different usage conditions, which can achieve a dual adjustment of the curing rate and the performance of the cured product [26].

3.1.3. Effect of Preparation Technology on the 60-Min Bond Strength

After filling the cracks with prepolymers containing 0.3%, 0.5%, 0.7%, and 1% mass fraction of the curing agent using the bonding strength test method described in 2.4.7, the repaired specimens were cured for a certain period of time and then subjected to a flexural strength test. The location of the fracture surface was observed, and the average flexural strength of three specimens for each dosage was taken as the final result of the bonding strength of the repair material, which was compared with the strength of mortar when broken. The test results are shown in Figure 9a–d, corresponding to different dosages of the curing agent.

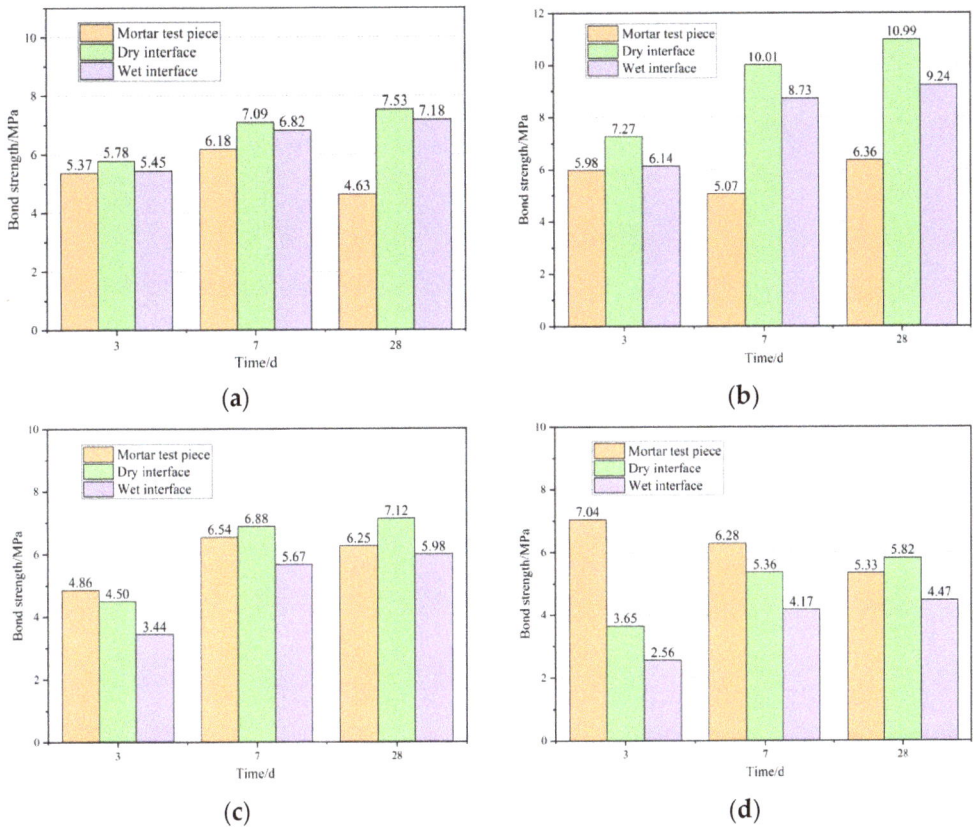

Figure 9. Relationship between wet and dry interface and bond strength: (**a**) 0.3% DMA; (**b**) 0.5% DMA; (**c**) 0.7% DMA; (**d**) 1% DMA.

It can be seen from Figure 9 that the bonding strength of the repair material at the dry interface and wet interface mostly exceeds the flexural strength of the mortar specimens themselves when broken. When the content of the curing agent is 0.3%, the bonding strength of the repair material at the dry interface and wet interface after 28 days is 7.53 Mpa and 7.18 Mpa, respectively. When the content of the curing agent is 0.5%, the peak bonding strength of the repair material at the dry interface and wet interface after 28 days is 10.99 Mpa and 9.24 Mpa, respectively. As the content of the curing agent increases, the bonding strength of the repair material gradually decreases. When the curing agent content is 1%, the bonding strength of the repair material on the dry interface and the wet interface after 28 days decreases to 5.82 Mpa and 4.47 Mpa, respectively, which is reduced by 47.0% and 51.6% compared with the 0.5% content. This indicates that with the increase in the curing agent content, the bonding strength of the repair material first reaches a peak and then gradually decreases.

The comparison of the bond strength of the repair material on the dry and wet interfaces shows that the bond strength on the wet interface is reduced by 3.8 to 29.9% compared to that on the dry interface. This indicates that the surface moisture of concrete directly affects the repair effect of the MMA-based repair materials. The presence of surface moisture affects the degree of curing of the material at the interface and reduces the roughness of the concrete interface, resulting in a decrease in the adhesion of the repair material to the concrete and a decrease in the bond strength. In contrast, on the dry interface, the penetrating repair material can penetrate the concrete to a certain depth, and the acrylic resin molecules can wrap and anchor the concrete, thereby enhancing the mechanical strength of the concrete to a certain extent [27].

3.2. Optimization of the Composition for the Penetrating Repair Material

Based on the optimal preparation process of MMA-based repair materials determined in Section 3.1, the average viscosity, drying time, and bonding properties of the repair materials with a DMA content of 0.3%, 0.5%, 0.7%, and 1% were comprehensively tested to further elucidate the variation law of material properties with DMA content. To ensure good infiltration performance of the repair material, a diluent was added to reduce the initial viscosity of the material, allowing it to penetrate the concrete surface to a certain depth without participating in the curing reaction.

3.2.1. Penetration Depth

To determine the trend of permeability of the MMA-based repair materials under different diluents, the effect of diluent type and content on the permeability of the repair materials was analyzed based on the determined optimal preparation process. The results are shown in Figure 10.

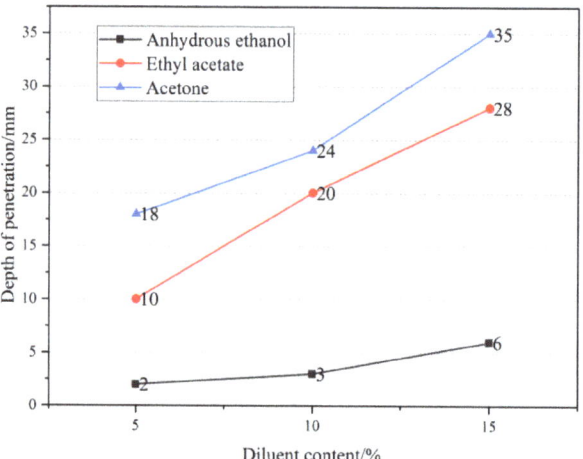

Figure 10. The influence of different diluent contents on the penetration depth.

As shown in Figure 10, the penetration depth of the three diluents increased with the increase in their content. The penetration depth of the three diluents at 5% content ranged from 2–18 mm, while at 10% content, it ranged from 3–24 mm. When the content increased to 15%, the penetration depth reached 6–35 mm. This indicates that increasing the diluent content can increase the penetration depth of the repair material. The presence of the diluent reduces the concentration of the system, and the viscosity of acetone at room temperature is 0.31 mPa·s, which significantly reduces the viscosity of the repair material system, enabling it to penetrate the concrete surface layer up to 35 mm under specified conditions. The viscosity of anhydrous ethanol at room temperature is 1.17 mPa·s. Due to the hydrophobic interaction between PMMA and anhydrous ethanol, when they are mixed, non-covalent binding occurs between PMMA particles, resulting in a white flocculent matter that severely affects the penetration performance of the repair material. In the mixture, PMMA is absorbed and forms white crystalline PMMA. Generally, at the same temperature, the initial viscosity has a critical impact on the infiltration performance of the material. Therefore, repair materials containing acetone and ethyl acetate can infiltrate

10–35 mm deep into the concrete, effectively sealing the pore structure on the concrete surface and improving its mechanical properties. Figure 11a–c shows the penetration of the repair materials with different diluents in concrete and the cross-sectional laser microscope images. It can be seen that the repair materials exist in the microcracks and pores of the concrete, penetrate to a certain depth on the surface, and form a high-performance solid structure, forming a "rooted type" filling.

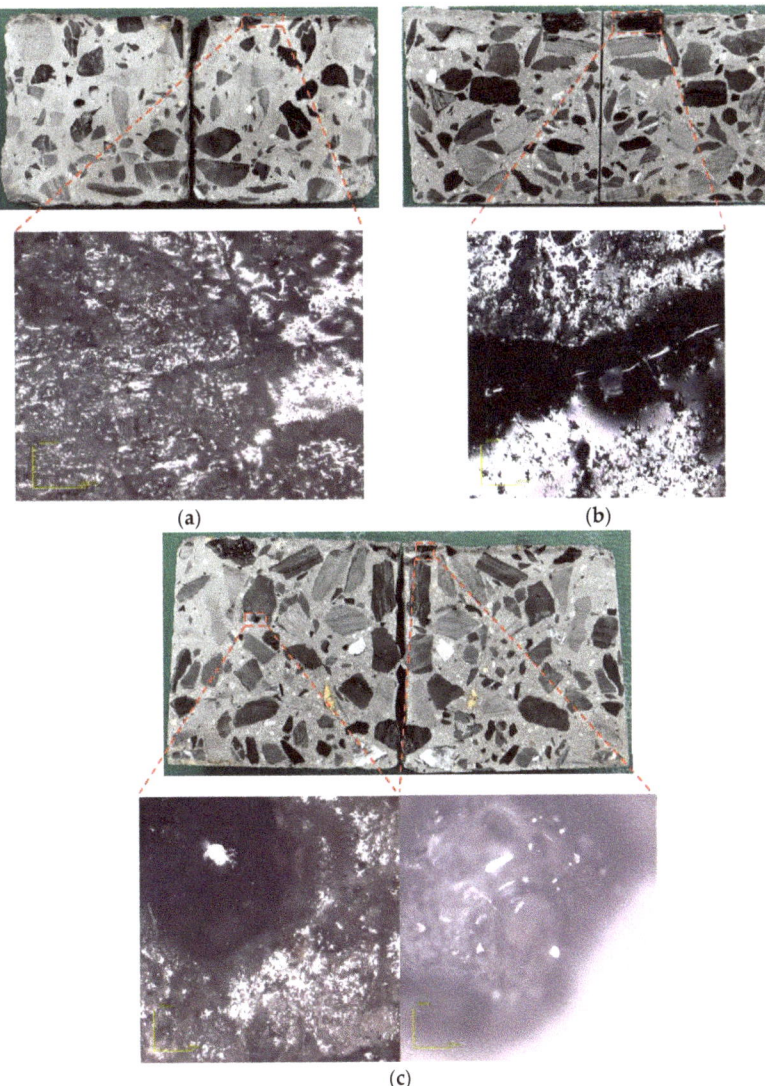

Figure 11. Infiltration condition: (**a**) anhydrous ethanol; (**b**) ethyl acetate; (**c**) acetone.

3.2.2. Surface Film Thickness

In order to elucidate the effect of the three diluents on the film thickness of the repaired concrete surface, blank samples and infiltrated solidified repair materials with different proportions were prepared. The film thickness of the repair material on the concrete surface was measured using laser confocal microscopy at 6 h, 24 h, and 72 h after preparation. The samples that were not completely cured and those that were fully cured were compared. The results are shown in Figures 12a–d and 13.

From Figure 12a–d, it can be seen that with the increase in time, the thickness of the repair material added with three different diluents showed a trend of rapid decrease followed by a slow decrease during the curing process. The thickness of W10, E10, and A10 after 72 h were 97 μm, 35 μm, and 26 μm, respectively, which decreased by 46.7%, 80.8%, and 85.7% compared to the repair material without diluent. The results showed that the addition of acetone resulted in the most significant reduction in film thickness, and the repair materials with diluents were able to penetrate the surface of the concrete well, reducing the film thickness on the surface and significantly reducing the impact on the excellent friction performance of the concrete surface. The fundamental reason for this lies in the low initial viscosity and high surface tension of the repair material in the first stage. Capillary action occurs between the material and the concrete capillary pores, and the additional pressure generated by the capillary action provides the driving force for penetration during the penetration process [28]. The greater the surface tension of the material and the smaller the contact angle between the material and the concrete, the more favorable it is for the material to penetrate the concrete structure [29]. In the second stage, as the oxidation–reduction reaction of the material continues, the already penetrated material gradually solidifies and forms a "rooted type" filler, while the material that has not penetrated accumulates and solidifies on the concrete surface to form a polymer film.

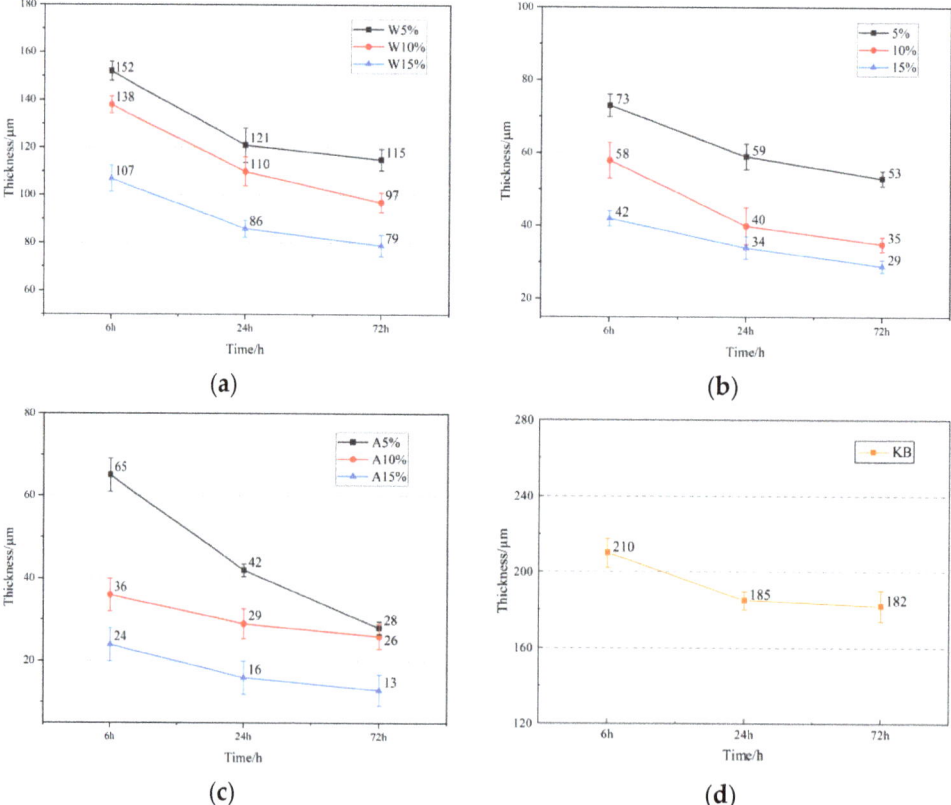

Figure 12. Effect of different diluents on film thickness: (**a**) anhydrous ethanol; (**b**) ethyl acetate; (**c**) acetone; (**d**) control.

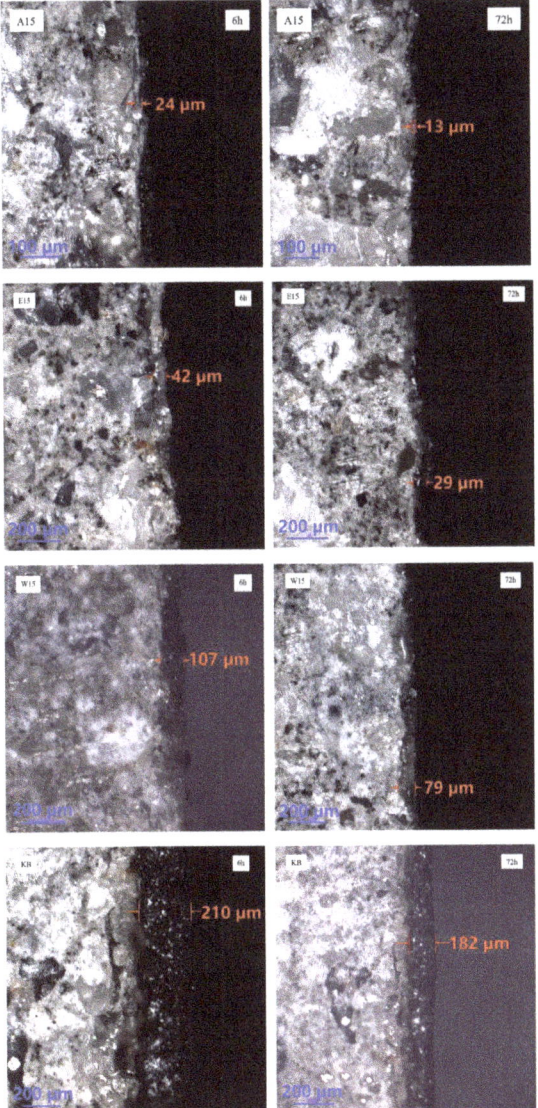

Figure 13. Results observed under a laser confocal microscope.

Figure 13 shows the comparison photos of thickness taken by laser confocal microscope. It can be clearly seen that the thickness of W15, E15, A15, and KB after 72 h are 79 μm, 29 μm, 13 μm, and 182 μm, respectively, which are decreased by 26.2%, 30.9%, 45.8%, and 13.3% compared to the thickness after 6 h. The results show that the addition of acetone in the repair materials leads to a significant decrease in the final film thickness with time. This is because acetone has a low viscosity, greatly reducing the initial viscosity of the repair material, expanding its surface tension, and providing a large amount of driving force for the infiltration process.

3.2.3. Freeze–Thaw Cycling Test

The influence of the freeze–thaw cycles on the repaired concrete was investigated. By simulating temperature changes, the protective effect of the repair materials on the concrete was tested. The final results were converted into the amount of peeling per square meter as the evaluation standard for the concrete's resistance to the freeze–thaw cycles. The results are shown in Figure 14a,b.

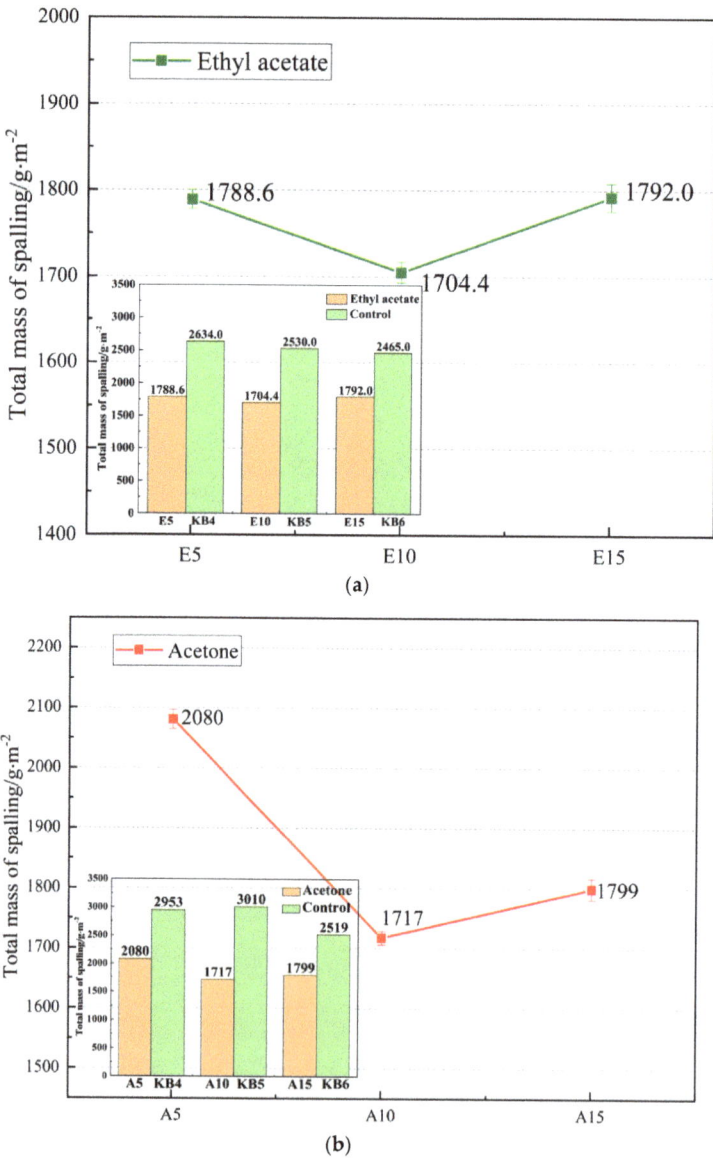

Figure 14. Effect of diluent content on freeze–thaw cycle: (**a**) effect of different contents of ethyl acetate on spalling amount; (**b**) effect of different acetone contents on spalling amount.

As shown in Figure 14a,b, it can be seen that the repair material added with 10% ethyl acetate has the best protective performance on concrete, and the average spalling amount per unit area is only 1704.4 g. Adding 5% acetone repair material for concrete protection performance is the worst, with an average spalling amount per unit area of 2080.8 g. Whether it is ethyl acetate or acetone, when the content is 5%, the amount of spalling shows a trend of slow and then fast, and when the content is 15%, the overall spalling speed is faster, which is because the mechanical strength of the repair material itself is significantly reduced, even if the penetration performance is excellent it cannot be well bonded and solidified with concrete, so the amount of solvent added should be between 10–15%.

3.2.4. SEM

Samples W10, E10, and A10 were applied to the surface of the concrete and cured for 28 days under natural conditions. Samples were taken from the surface of the broken parts and observed under SEM to investigate the bonding between the repair materials and cement paste. The results are shown in Figure 15.

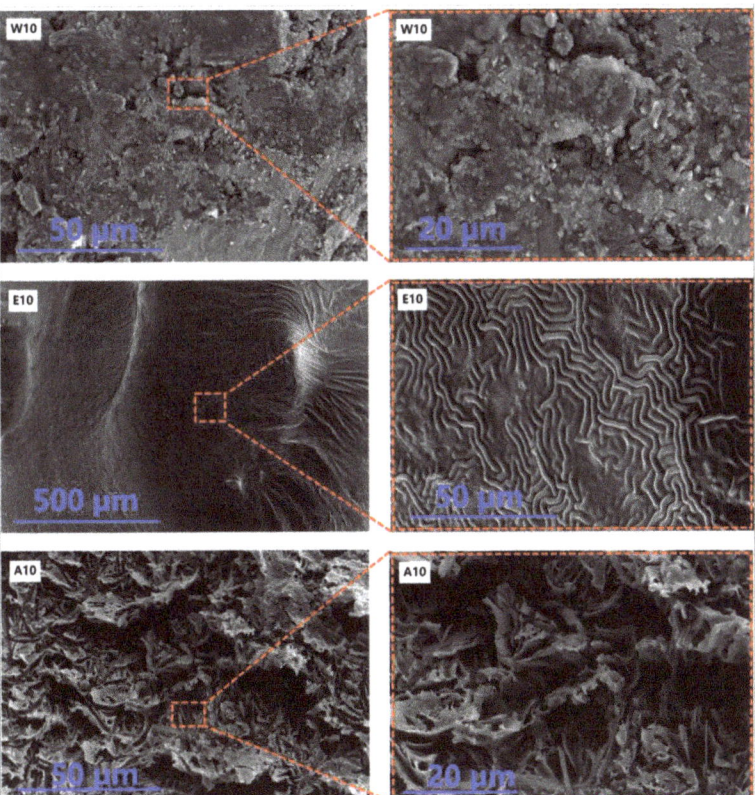

Figure 15. SEM images of three types of diluents.

Figure 15 shows the microstructure of the three samples. In the W10 sample, the repair material mostly remains on the surface of the cement paste, and the mortar surface is very rough and contains a large number of micropores, which provide channels for harmful substances to penetrate the interior of the concrete and do not provide effective protection. For the E10 sample, a polymer film was clearly visible on the surface of the mortar, and the repair material formed a tightly interwoven structure, which is conducive to preventing or delaying the penetration of corrosive media in the mortar. In contrast to the previous two samples, the A10 sample showed pinholes due to the rapid volatilization of solvents [30]. This is because when the repair material is dried, the volatilized solvent will form bubbles on the surface of the material. If the bubbles cannot be eliminated in time, pinhole-like small holes will be left on the surface of the coating.

4. Conclusions

1. In general, the optimum amount of curing agent is determined by adjusting the content of the curing agent. Then, by adding diluent, the repair material has good infiltration performance and can be consolidated in a certain depth of concrete surface, forming a "rooted type" filling so that the mechanical strength and durability of the concrete surface can be improved.
2. It was found that the 60-min average viscosity of the repair material increased first and then decreased as the content of DMA added was 0.3%, 0.5%, 0.7%, and 1%. When the reaction time reached 40–60 min, the viscosity of the system increased significantly. The viscosity of the repair material with 1% DMA content reached 284 mPa·s at 60 min, which was 46.12% higher than that of 0.3% DMA on average. Considering the controllability of viscosity and the prevention of explosion, DMA content in the range of 0.3–0.5% is more suitable.
3. After conducting surface-drying time, through-drying time, and bonding strength tests, it was found that the surface-drying time and through-drying time of the repair material with 1% DMA content were 0.8 h and 1.5 h, respectively, which were 60% and 67% shorter than those of the repair material with 0.3% DMA content. However, excessive addition of the curing agent can lead to a sharp increase in the reaction of the repair material, resulting in the formation of tiny bubbles. When the curing agent content was 0.5%, the bonding strength reached its peak, with values of 10.99 MPa and 9.24 MPa at the dry interface and wet interface, respectively. Therefore, the optimal curing agent content was determined to be 0.5%.
4. A comparison study of three different diluents showed that sample E10 had a penetration depth of 24 mm and a film thickness of only 35 μm after 72 h. After the freeze–thaw cycling test, E10 showed the best protective performance for concrete, with an average peeling strength of only 1704.4 g/m^2. SEM showed that the polymer formed a tightly interwoven protective film on the surface of the cement paste, effectively preventing the ingress of corrosive substances.

Author Contributions: Conceptualization, M.D.; data curation, P.L. and X.H.; writing—original. draft preparation, P.L.; writing—review and editing, M.D. and Z.M. All authors have read and agreed to the published version of the manuscript.

Funding: This work was supported by the Open Fund project of the National Laboratory for High Performance Civil Engineering Materials (2021CEM012) and the Priority Academic Program Development of Jiangsu Higher Education Institutions (PAPD).

Institutional Review Board Statement: Not applicable.

Informed Consent Statement: Not applicable.

Data Availability Statement: The data presented in this study are available on request from the corresponding author.

Acknowledgments: The authors gratefully acknowledge the assistance from Zhongyang Mao and Xiaojun Huang from NJTECH and the staff from the State Key Laboratory of Materials-Oriented Chemical Engineering.

Conflicts of Interest: The authors declare no conflict of interest.

References

1. Si-pei, L.I. Some Common Quality Defects on Cement Concrete Pavement and Precautions. *Sci-Tech Inf. Dev. Econ.* **2009**, *9*, 230–232.
2. Berrocal, C.G.; Löfgren, I.; Lundgren, K.; Tang, L. Corrosion initiation in cracked fibre reinforced concrete: Influence of crack width, fibre type and loading conditions. *Corros. Sci.* **2015**, *98*, 128–139. [CrossRef]
3. Basheer, L.; Kropp, J.; Cleland, D.J. Assessment of the durability of concrete from its permeation properties: A review. *Constr. Build. Mater.* **2001**, *15*, 93–103. [CrossRef]
4. Flores-Vivian, I.; Hejazi, V.; Kozhukhova, M.I.; Nosonovsky, M.; Sobolev, K. Self-Assembling Particle-Siloxane Coatings for Superhydrophobic Concrete. *ACS Appl. Mater. Inter.* **2013**, *5*, 13284–13294. [CrossRef]
5. Fardin, H.E.; Santos, A.G.D. Predicted responses of fatigue cracking and rutting on Roller Compacted Concrete base composite pavements. *Constr. Build. Mater.* **2021**, *272*, 121847. [CrossRef]
6. Xiao, Y.; Erkens, S.; Li, M.; Ma, T.; Liu, X. Sustainable Designed Pavement Materials. *Materials* **2020**, *13*, 1575. [CrossRef]
7. Litina, C.; Al-Tabbaa, A. First generation microcapsule-based self-healing cementitious construction repair materials. *Constr. Build. Mater.* **2020**, *255*, 119389. [CrossRef]
8. Cabrera, J.G.; Al-Hasan, A.S. Performance Properties of Concrete Repair Materials. *Constr. Build. Mater.* **1997**, *11*, 283–290. [CrossRef]
9. Venkiteela, G.; Klein, M.; Najm, H.; Balaguru, P.N. Evaluation of the Compatibility of Repair Materials for Concrete Structures. *Int. J. Concr. Struct. Mater.* **2017**, *11*, 435–445. [CrossRef]
10. Al-kahtani, M.S.M.; Zhu, H.; Ibrahim, Y.E.; Haruna, S.I. Experimental study on the strength and durability-related properties of ordinary Portland and rapid hardening Portland cement mortar containing polyurethane binder. *Case Stud. Constr. Mater.* **2022**, *17*, e1530. [CrossRef]
11. Feng, H.; Lu, L.; Chen, J.; Wang, G.; Zhang, Y. Influence of fly ash/silica fume composite admixtures on the performance of cement paste. *Appl. Chem. Ind.* **2014**, *43*, 389–391, 394.
12. Wu, R.; Wang, X. Progress in Polyurethane Based Road Repair Materials. *J. Mater. Sci. Eng.* **2021**, *39*, 163, 164–166.
13. Lee, S. Mechanical Properties and Durability of Mortars Made with Organic-Inorganic Repair Material. *J. Test. Eval.* **2021**, *49*, 4378–4389. [CrossRef]
14. Zhang, G.; Zhu, Z.; Ma, C.; Zhang, G. Organic/inorganic dual network formed by epoxy and cement. *Polym. Compos.* **2018**, *39*, E2490–E2496. [CrossRef]
15. Kim, T.K.; Park, J.S. Experimental Evaluation of the Durability of Concrete Repair Materials. *Appl. Sci.* **2021**, *11*, 2303. [CrossRef]
16. Hassan, K.E.; Brooks, J.J.; Al-Alawi, L. Compatibility of repair mortars with concrete in a hot-dry environment. *Cem. Concr. Compos.* **2001**, *23*, 93–101. [CrossRef]
17. Khalina, M.; Beheshty, M.H.; Salimi, A. The effect of reactive diluent on mechanical properties and microstructure of epoxy resins. *Polym. Bull.* **2019**, *76*, 3905–3927. [CrossRef]
18. Pang, B.; Jin, Z.; Zhang, Y.; Xu, L.; Li, M.; Wang, C.; Zhang, Y.; Yang, Y.; Zhao, P.; Bi, J.; et al. Ultraductile waterborne epoxy-concrete composite repair material: Epoxy-fiber synergistic effect on flexural and tensile performance. *Cem. Concr. Compos.* **2022**, *129*, 104463. [CrossRef]
19. Li, X.; Hao, M.; Zhong, Y.; Zhang, B.; Wang, F.; Wang, L. Experimental study on the diffusion characteristics of polyurethane grout in a fracture. *Constr. Build. Mater.* **2021**, *273*, 121711. [CrossRef]
20. Wang, Z.Q.; Wei, T.; Li, Z.; Jiang, S.Z.; Xue, X.L. Study and Application of CW Epoxy Resin Chemical Grouting Materials. *J. Yangtze River Sci. Res. Inst.* **2011**, *28*, 167–170.
21. Issa, C.A.; Debs, P. Experimental study of epoxy repairing of cracks in concrete. *Constr. Build. Mater.* **2007**, *21*, 157–163. [CrossRef]
22. Safan, M.A.; Etman, Z.A.; Konswa, A. Evaluation of polyurethane resin injection for concrete leak repair. *Case Stud. Constr. Mater.* **2019**, *11*, e307. [CrossRef]
23. Han, J.; Xu, L.; Feng, T.; Shi, X.; Zhang, P. Effect of PCE on Properties of MMA-Based Repair Material for Concrete. *Materials* **2021**, *14*, 859. [CrossRef] [PubMed]
24. Sabah, S.A.; Hassan, M.H.; Bunnori, N.M.; Johari, M.M. Bond strength of the interface between normal concrete substrate and GUSMRC repair material overlay. *Constr. Build. Mater.* **2019**, *216*, 261–271. [CrossRef]
25. Besheli, A.E.; Samimi, K.; Nejad, F.M.; Darvishan, E. Improving concrete pavement performance in relation to combined effects of freeze–thaw cycles and de-icing salt. *Constr. Build. Mater.* **2021**, *277*, 122273. [CrossRef]
26. Pojman, J.A. Cure-on-Demand Composites by Frontal Polymerization. In *Encyclopedia of Materials: Plastics and Polymers*; Hashmi, M.S.J., Ed.; Elsevier: Oxford, UK, 2022; pp. 85–100. ISBN 978-0-12-823291-0.
27. Cui, Y.; Tan, Z.; An, C. Research and application of multi-functional acrylic resin grouting material. *Constr. Build. Mater.* **2022**, *359*, 129381. [CrossRef]
28. Looney, T.; Leggs, M.; Volz, J.; Floyd, R. Durability and corrosion resistance of ultra-high performance concretes for repair. *Constr. Build. Mater.* **2022**, *345*, 128238. [CrossRef]

29. Krainer, S.; Hirn, U. Contact angle measurement on porous substrates: Effect of liquid absorption and drop size. *Colloids Surf. A Physicochem. Eng. Asp.* **2021**, *619*, 126503. [CrossRef]
30. Fitzsimons, B.; Parry, T. Paint and Coating Failures and Defects. In *Reference Module in Materials Science and Materials Engineering*; Elsevier: Amsterdam, The Netherlands, 2016; ISBN 978-0-12-803581-8.

Disclaimer/Publisher's Note: The statements, opinions and data contained in all publications are solely those of the individual author(s) and contributor(s) and not of MDPI and/or the editor(s). MDPI and/or the editor(s) disclaim responsibility for any injury to people or property resulting from any ideas, methods, instructions or products referred to in the content.

Article

Effect of Different Expansive Agents on the Deformation Properties of Core Concrete in a Steel Tube with a Harsh Temperature History

Anqun Lu [1,2,*], Wen Xu [3], Qianqian Wang [4,*], Rui Wang [5] and Zhiyuan Ye [1,2]

1. State Key Laboratory of High Performance Civil Engineering Materials, Jiangsu Research Institute of Building Science, Nanjing 210008, China
2. Research Institute of Jiangsu Sobute New Materials Co., Ltd., Nanjing 211103, China
3. School of Materials Science and Engineering, Southeast University, Nanjing 211189, China
4. College of Materials Science and Engineering, Nanjing Tech University, Nanjing 210009, China
5. Lalin Railway Construction Headquarters of China Railway Corporation, Linzhi 860114, China
* Correspondence: luanqun@cnjsjk.cn (A.L.); qqwang@njtech.edu.cn (Q.W.)

Abstract: The shrinkage of core concrete during construction is the key reason for the separation of steel pipes and core concrete. Utilizing expansive agents during cement hydration is one of the main techniques to prevent voids between steel pipes and core concrete and increase the structural stability of concrete-filled steel tubes. The expansion and hydration properties of CaO, MgO, and CaO + MgO composite expansive agents in C60 concrete under variable temperature conditions were investigated. The effects of the calcium–magnesium ratio and magnesium oxide activity on deformation are the main parameters to consider when designing composite expansive agents. The results showed that the expansion effect of CaO expansive agents was predominant in the heating stage (from 20.0 °C to 72.0 °C at 3 °C/h), while there was no expansion in the cooling stage (from 72.0 °C to 30.0 °C at 3 °C/d, and then to 20.0 °C at 0.7 °C/h); the expansion deformation in the cooling stage was mainly caused by the MgO expansive agent. With the increase in the active reaction time of MgO, the hydration of MgO in the heating stage of concrete decreased, and the expansion of MgO in the cooling stage increased. During the cooling stage, 120 s MgO and 220 s MgO resulted in continuous expansion, and the expansion curve did not converge, while 65 s MgO reacted with water to form brucite in large amounts, leading to its lower expansion deformation during the later cooling process. In summary, the CaO and 220 s MgO composite expansive agent in the appropriate dosage is suitable for compensating for the shrinkage of concrete in the case of a fast high-temperature rise and slow cooling rate. This work will guide the application of different types of CaO-MgO composite expansive agents in concrete-filled steel tube structures under harsh environmental conditions.

Keywords: CaO and MgO composite expansive agent; reaction time of MgO; temperature history; expansion properties

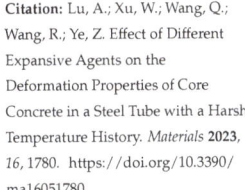

Citation: Lu, A.; Xu, W.; Wang, Q.; Wang, R.; Ye, Z. Effect of Different Expansive Agents on the Deformation Properties of Core Concrete in a Steel Tube with a Harsh Temperature History. *Materials* **2023**, *16*, 1780. https://doi.org/10.3390/ma16051780

Academic Editor: Weiting Xu

Received: 17 January 2023
Revised: 12 February 2023
Accepted: 16 February 2023
Published: 21 February 2023

Copyright: © 2023 by the authors. Licensee MDPI, Basel, Switzerland. This article is an open access article distributed under the terms and conditions of the Creative Commons Attribution (CC BY) license (https://creativecommons.org/licenses/by/4.0/).

1. Introduction

Concrete-filled steel tubes, as composite structures, can enhance the performance of concrete and steel. It has a higher bearing capacity, stability, and economic benefit when compared to the individual components; hence, it is frequently utilized in large bridge structures [1–3]. Materials scientists and structural engineers have made great progress on its wide application, and hundreds of concrete-filled steel tube arch bridges have been built or are under construction worldwide, which has resulted in a wide range of structural forms and significant progress in the building of long-span arch bridges [4–7]. However, in the construction of concrete-filled steel tubes, the problems of interface voids and debonding are common, affecting the synergy between them and endangering structural safety [8,9]. The debonding and voiding of concrete-filled steel tubes happen for a variety of reasons, including axial compression, shrinkage and creep of the concrete itself, and internal and exterior temperature variations caused by sunlight [10–12]. The interface void problem occurs once the construction period of most arch bridges is finished, indicating that the shrinkage of concrete during the construction period is the primary reason for voids [13]. The core concrete in the steel tube has a high strength grade in general, and it experiences a severe temperature drop during its early hydration period. Uneven internal and exterior deformations will cause the concrete surface to debond from the steel. In the later stage, the superposition of concrete temperature-drop-induced shrinkage and autogenous shrinkage will further aggravate the void problem. Therefore, controlling the shrinkage of concrete in concrete-filled steel tubes has emerged as an important topic [14].

Decreasing the shrinkage or causing the slight expansion of the concrete in the steel pipe could be crucial to preventing the separation of the steel pipe and concrete. Utilizing expansive agents during cement hydration is one of the main techniques to prevent voids between the steel pipe and core concrete and increase the structural stability of concrete-filled steel tubes [15]. In cement and concrete, three types of expansive agents are commonly used: sulfoaluminate hydrated ettringite [16–18], calcium oxide (CaO) hydrated calcium hydroxide [19,20], and magnesium oxide (MgO) hydrated magnesium hydroxide [21–23]. The classic calcium sulfoaluminate expansive agent has the disadvantages of high water demand and unstable hydration products at high temperatures, limiting its usage in high-strength concrete with a lower water–binder ratio [24,25]. The calcium oxide expansive agent is widely used in concrete-filled steel tube structures [26], but its hydration speed is too fast, the adjustability of the expansion process is poor, and its hydration is largely ineffective before the formation of the concrete slurry aggregate structure (that is, the plastic stage) [27]. According to previous studies [28,29], the compensatory impact of calcium oxide expansive agents on the cooling shrinkage and drying shrinkage of high-strength concrete is insignificant. Compared with the CaO expansive agent, the hydration products of the MgO expansive agent are more stable, and its expansion process can be much easier to control [30,31]. The MgO expansive agent has been widely used to compensate for the cooling shrinkage and autogenous shrinkage of hydraulic mass concrete [32,33]. Yao et al. [34] and Cai et al. [35] examined the deformation properties of a microexpansive concrete-filled steel tube with a MgO-based expansive agent at room temperature.

Since the temperature of concrete changes during the construction process, the performance of concrete at room temperature cannot reflect its application performance in the actual building. For example, for a steel tube filled with C50~C80 concrete with a large pipe diameter, the central temperature of core concrete can reach 50~60 °C. Due to heat dissipation to the environment, the concrete will undergo a rapid temperature drop after reaching the temperature peak. For example, during the construction of the Zangmu Bridge in Tibetan areas in China, the average temperature drop rate of the core concrete reached 3 °C/d after the temperature rise, which created a big challenge and required building non-void concrete-filled steel tubes. Thus, studying the effect of an expansive agent exposed to the actual temperature history is more valuable for engineering applications.

Furthermore, utilizing a single type of expansive agent to compensate for the shrinkage deformation of concrete in different stages is not sufficient. Liu et al. [36] designed a magnesium oxide composite expansion agent composed of specific proportions of high-activity MgO, low-activity MgO, and CaO at 20 °C, which effectively eliminated the early autogenous shrinkage of high-performance concrete and markedly inhibited its drying shrinkage. Yu et al. [37] also explored a type of multisource expansive agent mixed with different proportions of calcium oxide, calcium sulfoaluminate, and MgO in high-strength concrete, which revealed that the multisource expansive agent can effectively compensate for the drying shrinkage of high-strength concrete at normal temperatures in the later stage. Thus, previous research on cement-based materials mixed with CaO and MgO composite expansive agents has mainly focused on strength and deformation development at constant temperatures. Recently, Zhao et al. [38] studied the effect of a blended MgO-CaO expansive agent on the hydration of cement paste at an early age by using low-field nuclear magnetic resonance technology, and they found that the curing temperature has a great impact on the hydration of the cement paste mixed with the MgO-CaO blended expansive agent. However, the influence of the activity of MgO in the CaO and MgO composite expansive agent on the deformation properties has not been studied. Additionally, there are few studies on the influence of the calcium-to-magnesium ratio of CaO and MgO composite expansive agents on the deformation of concrete when they are under variable temperature conditions during the construction process.

This work investigates the deformation performance of C60 concrete mixed with different contents of a CaO expansive agent, MgO expansive agent, and CaO-MgO composite expansive agents under the condition of a simulated temperature history. The temperature variation process used in this work was first monitored for a part of the core concrete in the steel tubes of the Zangmu Bridge in Tibet in China. This temperature history could be suitable for simulating the deformation performance of concrete under the typically closed condition of the harsh plateau environment. Afterward, a reasonable amount of a calcium oxide and magnesium oxide composite expansive agent was mixed with concrete and poured into tubes. The effect of the CaO-MgO ratio of the composite expansive agent on the concrete was investigated by testing its deformation performance by using an SBT-CDMI wireless monitoring system, and the activity time of MgO on the deformation was considered. Finally, the hydration products and microstructures of cement pastes with different types of expansive agents were also studied. These findings will guide the use of CaO-MgO composite expansive agents in concrete-filled steel tubes and other non-shrink concrete.

2. Materials and Methods

2.1. Materials

The cement used in this study was ordinary Portland cement 42.5, which was produced by Conch Cement Plant, and conformed to the Chinese standard GB175-2007 [39]. Fly ash was Class II fly ash from the Huaneng Power Plant. Natural river sand with a fineness modulus of 2.60 was used. The aggregate was basalt gravel. The particle size of small stones was 5–20 mm. The particle size of medium stones was 20–40 mm. The superplasticizer was PCA polycarboxylic acid superplasticizer produced by Jiangsu Sobute New Materials Co., Ltd., Nanjing, China.

Three MgO expansive agents with different active reaction times (65 s, 120 s, and 220 s) were prepared by calcining magnesite in a suspension kiln. The 65 s, 120 s, and 220 s MgO expansive agents were calcined from magnesite powder (particle size $\leq$ 160 μm) at 800 °C, 950 °C, and 1050 °C, respectively, in a suspension kiln. The citric acid method was used to determine the active reaction time of the MgO expansive agent [40]. The active reaction time was determined by using 1.70 ± 0.1 g of the MgO expansive agent to completely neutralize 200 mL of a 0.07 mol/L citric acid solution at 30 ± 1 °C, which was used as a measure for evaluating the activity of the MgO expansive agent. Obviously, a shorter active reaction time means higher activity [30]. The CaO expansive agent was prepared by calcining limestone and a mineralizer in a rotary kiln at 1350 °C. Both the MgO expansive agent and CaO expansive agent were provided by Jiangsu Sobute New Materials Co., Ltd., Nanjing, China. The chemical components of cement, the MgO expansive agent, and the CaO expansive agent are demonstrated in Table 1.

The XRD patterns of MgO expansive agents and CaO expansive agents are shown in Figure 1. The 65 s MgO, 120 s MgO, and 220 s MgO expansive agents have MgO as the main mineral and contain small amounts of SiO_2 and CaO. The primary mineral of the CaO expansive agent is CaO, containing a small amount of $CaSO_4$.

2.2. Experiments

2.2.1. Mix Design of Concrete

Table 2 displays the concrete proportion for the C60 strength grade. MgO and CaO expansive agents were mixed internally. The slump of the concrete was controlled by the superplasticizer and other admixtures at 180–200 mm with 3.0–5.0% air content. The table also shows the amount of fly ash in the design of C60 concrete. Single CaO expansive agents, single MgO expansive agents, and compound expansive agents in C60 concrete were added to the mixture. It can be seen from the table that 6% CaO + 2% 65 s MgO indicates that the 6% calcium oxide expansive agent and the 2% magnesium oxide expansive agent with an active reaction time of 65 s are mixed in the concrete.

Table 1. Chemical compositions of cement, CaO expansive agent, and MgO expansive agents with different reaction times.

Raw Material	Reaction Times (s)	Chemical Composition (wt%)						
		MgO	CaO	SO_3	SiO_2	Al_2O_3	Fe_2O_3	Loss
Conch Cement	-	1.35	60.05	3.35	23.60	3.95	5.43	1.15
65 s MgO	65	90.17	1.87	-	1.83	1.39	2.15	2.59
120 s MgO	120	90.93	1.96	-	1.71	1.24	2.00	2.16
220 s MgO	220	91.38	2.24	-	2.36	1.32	1.12	1.58
CaO expansive agent	-	1.27	87.5	3.41	1.72	4.79	3.78	0.94

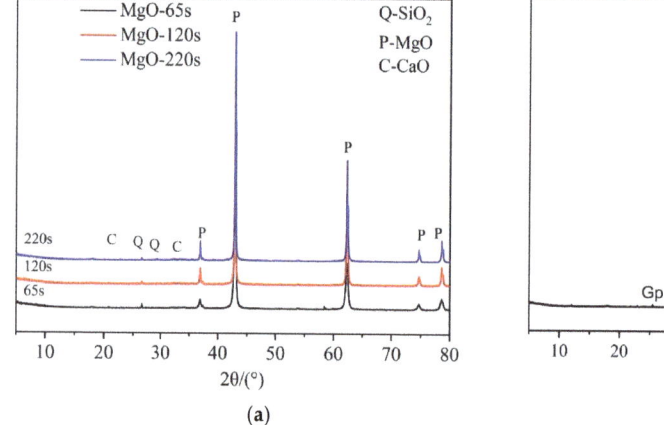
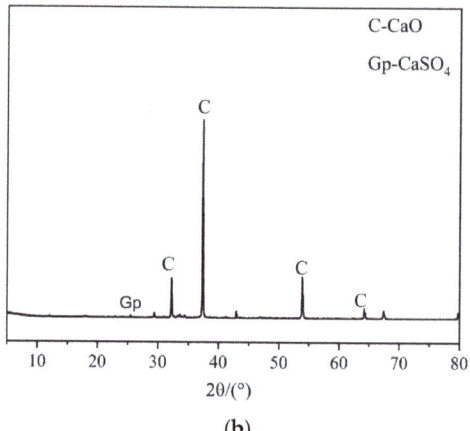

Figure 1. XRD patterns of expansive agents: (**a**) MgO expansive agents and (**b**) CaO expansive agents.

Table 2. Concrete mix designs of C60.

No.	W/C	Mix Ratio (kg/m³)							
		Cement	Fly Ash	CaO Expansive Agent	MgO Expansive Agent	Water	Sand	Small Basalt	Medium Basalt
C60-ref	0.29	416	104	0	0	151	720	249	746
6% CaO + 2% 65 s MgO	0.29	416	62.4	31.2	10.4	151	720	249	746
6% CaO + 2% 120 s MgO	0.29	416	62.4	31.2	10.4	151	720	249	746
6% CaO + 2% 220 s MgO	0.29	416	62.4	31.2	10.4	151	720	249	746
6% CaO + 4% 65 s MgO	0.29	416	52	31.2	20.8	151	720	249	746
6% CaO	0.29	416	72.8	31.2	0	151	720	249	746
8% CaO	0.29	416	62.4	41.6	0	151	720	249	746
4% 65 s MgO	0.29	416	83.2	0	20.8	151	720	249	746
4% 120 s MgO	0.29	416	83.2	0	20.8	151	720	249	746

Replacing a portion of the cement with fly ash can reduce the amount of heat released in concrete, the adiabatic temperature rise in concrete, and the autogenous shrinkage of concrete, which decreases the risk of concrete cracking, particularly in concrete containing a large amount of fly ash. Therefore, fly ash was used in this design for its application in concrete-filled steel tubes. The added content of MgO was selected according to T/CECS 10082—2020 Calcium and Magnesium Oxides Based Expansive Agent for Concrete [41].

2.2.2. Test of Deformation Performance of C60 Concrete under Variable Temperature Conditions

The test was conducted using a temperature and humidity environment simulation test chamber for concrete made by Nanjing Huanke Testing Equipment Co., Ltd., Nanjing, China., to simulate the temperature history of a C60 concrete-filled steel tube. The temperature and humidity environment simulation test chamber for concrete is shown in Figure 2. First, fresh C60 concrete was poured into polyvinyl chloride (PVC for short) pipes (Φ160 mm × 400 mm, shown in Figure 3). Then, a temperature–strain sensor was embedded in the center of the PVC pipe that had been filled with concrete (shown in Figure 4). The upper part of the PVC pipe was sealed with tin foil. Finally, the PVC pipes were put into the test chamber to simulate the environment (shown in Figure 5). The experiments used a pore water pressure testing device to measure the development of pore water pressure in concrete to determine the setting time [42]. An SBT-CDMI wireless monitoring system for the concrete temperature and strain was used to monitor the concrete's deformation and temperature history. The wireless monitoring system and the pore water pressure testing device were provided by Jiangsu Sobute New Materials Co. Ltd., Nanjing, China.

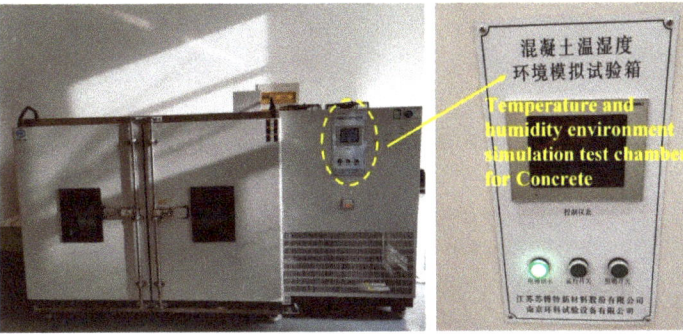

Figure 2. Temperature and humidity environment simulation test chamber for concrete.

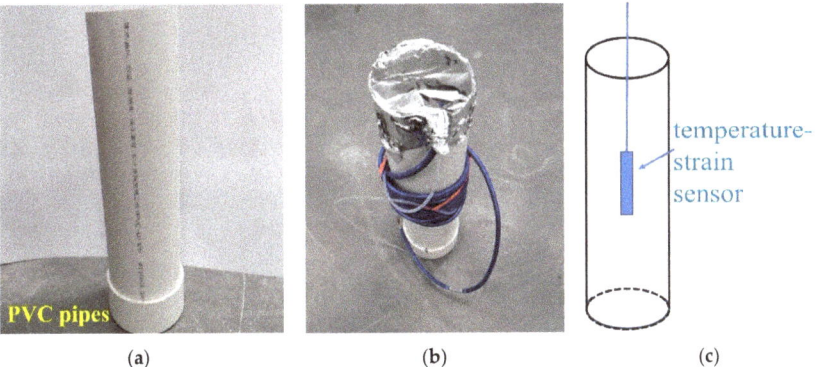

Figure 3. PVC pipe and PVC pipe filled with concrete and temperature–strain sensor embedded in the central part: (**a**) PVC pipe, (**b**) PVC pipe containing concrete, and (**c**) schematic diagram.

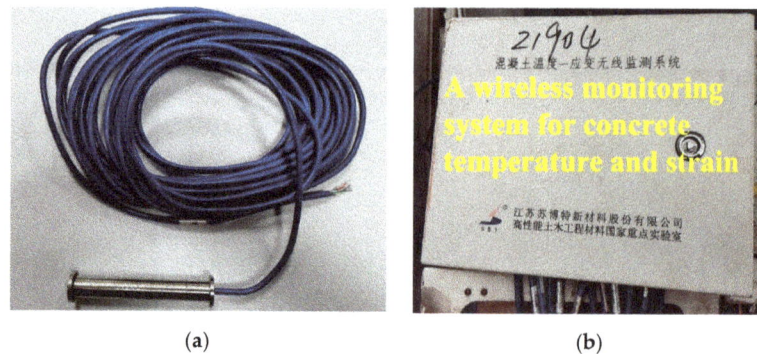

Figure 4. Monitoring by SBT-CDMI wireless monitoring system for concrete temperature and strain: (**a**) sensor and (**b**) data receiving system.

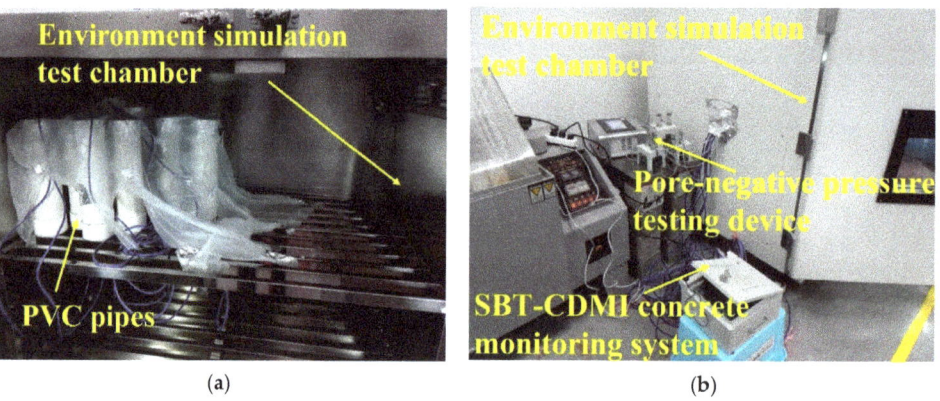

Figure 5. Concrete deformation and temperature monitoring test: (**a**) test molds and devices and (**b**) data receiving system.

The C60 concrete in the PVC pipe was in a state of absolute humidity. The concrete deformation recorded by the wireless monitoring system was mainly caused by autogenous volume deformation and temperature deformation. The temperature history of the concrete was mainly controlled by the environmental temperature change in the test chamber. Firstly, the test chamber was started, and the initial temperature inside the chamber was controlled at (20.0 ± 1.0) °C. After 20 h at a constant temperature, the test chamber was heated up to (72.0 ± 1.0) °C at 3 °C/h, then cooled down to (30.0 ± 1.0) °C at 3 °C/d, and then cooled down to (20.0 ± 1.0) °C at 0.7 °C/h. After the temperature fell to (20.0 ± 1.0) °C, the test chamber was turned off. The temperature of concrete varied with the external environment. The temperature variation process used in this work was first monitored for a part of the core concrete in the steel tubes of the Zangmu Bridge in Tibet in China. Its typical temperature history can be used to simulate the deformation performance of concrete embedded in large-diameter tubes under the closed condition of the harsh plateau environment.

2.2.3. Hydration Heat

Samples were placed at a constant temperature of 20 °C about 24 h before the experiment. Then, 10.0 g of the expansive agent was weighed and put into an ampoule bottle, and 10 mL of water was injected into the bottle with a syringe. Then, the paste was stirred quickly and evenly and put into the test channel of a TAM AIR isothermal calorimeter. The hydration and cumulative heat release rates were tested for 3 consecutive days.

2.2.4. Measurement of Hydration Degree of MgO Expansion Agent under Variable Temperature Conditions

To prevent the influence of components such as sand and gravel aggregate on the analysis of experimental results, the experiment was conducted with a cementitious material paste specimen. The expansive agent replaces the fly ash in an equal amount and does not replace the cement. The cement paste specimen was made according to a water–binder ratio of 0.29. After the paste sample was stirred evenly, it was poured into a 100 mL plastic test tube (shown in Figure 6). The plastic test tube and the above PVC pipes filled with concrete were put into the environmental simulation test chamber. After the variable-temperature experiment, the plastic test tube was taken out. The sample was put into an agate mortar, mixed, and ground with an appropriate amount of absolute ethanol until it passed through an 80 μm square sieve. The test instrument was a D8 Advance X-ray diffractometer manufactured by Bruker Company in Ettlingen, Germany, with a voltage of 40 kV, current of 200 mA, and scanning angle range of 5°~70°. The quantitative analysis of Rietveld full-spectrum fitting was performed with Jade 6 software.

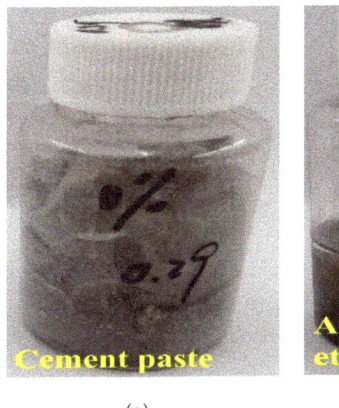
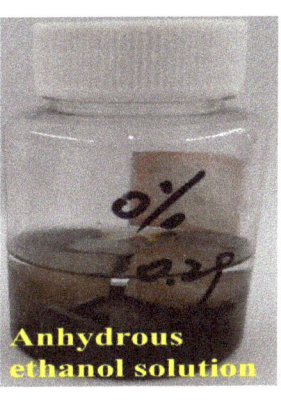

(a) (b)

Figure 6. Cementitious material paste specimen poured into a 100mL plastic test tube: (**a**) cement paste in a plastic test tube and (**b**) cement paste sample dehydrated in anhydrous ethanol solution.

2.2.5. SEM Analyses of Cement Paste Mixed with Expansive Agent

The micromorphology of cement paste with the expansive agent was characterized by SEM. The samples were processed as follows: cutting the two ends of the hardened paste in the plastic bottle to 5 mm, crushing the rest into small pieces, putting them into anhydrous ethanol for dehydration for 24 h (shown in Figure 6), drying them to constant weight, and sealing them for storage. A Quanta 250 field-emission scanning microscope manufactured by FEI Company in Hillsboro, OR, USA, was used to observe the micromorphology of a fresh fracture in the slurry. The fracture surface was presprayed with gold film.

3. Results and Discussion

3.1. Expansion Properties of CaO Expansive Agents in C60 Concrete under Variable Temperature Conditions

The autogenous volume deformation and temperature history of C60 concrete with 0%, 6%, and 8% CaO expansive agents under simulated variable temperature conditions are shown in Figure 7. The experiment used a pore water pressure testing device to measure the development of pore water pressure in concrete to determine the setting time. The time was taken as the initial setting time when the pore water pressure in concrete reached

10 kPa [42]. With the initial setting time as the zero point, the C60-ref reference concrete in the heating stage underwent 298 με expansion deformation, and the temperature rose to 52.5 °C. Moreover, C60 concrete with the 6% CaO expansive agent in the heating stage gave rise to 672 με expansion deformation, and the temperature rose to 53.1 °C. In addition, C60 concrete with the 8% CaO expansive agent in the heating stage triggered 761 με expansion deformation, and the temperature rose to 53.4 °C. C60 concrete containing the CaO expansive agent was still in an expansion state at 24.5 days after the initial setting time under variable temperature conditions.

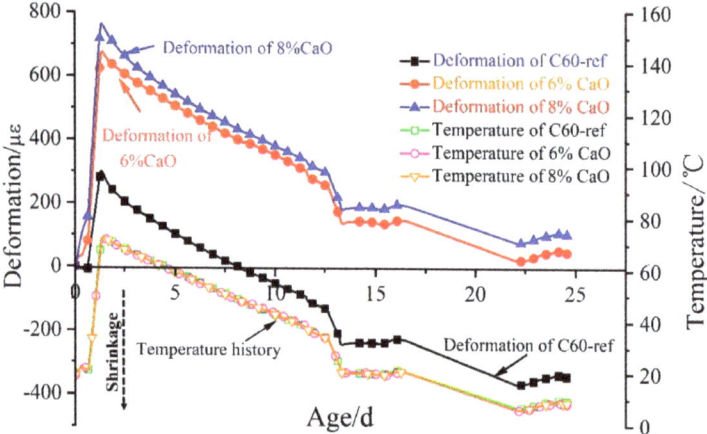

Figure 7. Deformation and temperature history of C60 concrete with 0%, 6%, and 8% CaO expansive agents.

The deformation process of C60 concrete with 0%, 6%, and 8% CaO expansive agents in the cooling stage is depicted in Figure 8. The shrinkage of C60-ref reference concrete was 637 με at 23 d of age in the cooling stage, while the shrinkage of C60 concrete with the 6% CaO expansive agent was 626 με. Furthermore, the shrinkage of C60 concrete with the 8% CaO expansive agent was 657 με. Among them, the shrinkage of C60 concrete with the 6% CaO expansive agent in the cooling phase was similar to that of C60-ref concrete, while the shrinkage of concrete with the 8% CaO expansive agent in the cooling phase was slightly larger than that of C60-ref concrete.

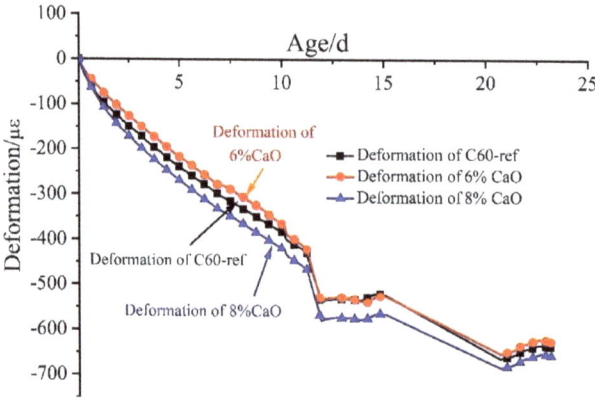

Figure 8. Deformation process of C60 concrete with 0%, 6%, and 8% CaO expansive agents in cooling stage.

The hydration of the CaO expansive agent in concrete to form Ca(OH)$_2$ crystals will result in the volume expansion of concrete, especially with higher amounts of the CaO expansive agent. Details are provided in Section 3.7.

Figure 9 illustrates the expansion deformation of 6% and 8% CaO expansive agents and the base concrete (C60-ref)'s net deformation effects, with the initial setting time as the zero point. The expansion of the CaO expansive agent was significant during the heating stage of the concrete. The concrete reached the temperature peak at 1.46 d after the initial setting time, and the 6% CaO expansive agent generated 382 με expansion deformation. Subsequently, the 6% CaO expansive agent generated 18 με expansion deformation during the cooling stage from 1.46 d to 3.00 d. At 24.5 d, the expansion produced by the 6% CaO expansive agent was 389 με. During the cooling stage from 1.46 d to 24.5 d, the expansion of the 6% CaO expansive agent increased by only 7 με. The concrete with the 8% CaO expansive agent reached the temperature peak at 1.46 d after the initial setting time, and the 8% CaO expansive agent generated 466 με expansion deformation. During the cooling stage from 1.46 d to 3.00 d, the expansion of the 8% CaO expansive agent was reduced by 17 με. From the initial setting time to 24.5 d, the 8% CaO expansive agent generated 445 με expansion deformation. During 24.5 days after the initial setting time, the expansion of the 6% CaO expansive agent in the heating stage accounted for 98.2% of the total expansion. The expansion of the 8% CaO expansive agent in the heating stage decreased by 4.5% at 24.5 d.

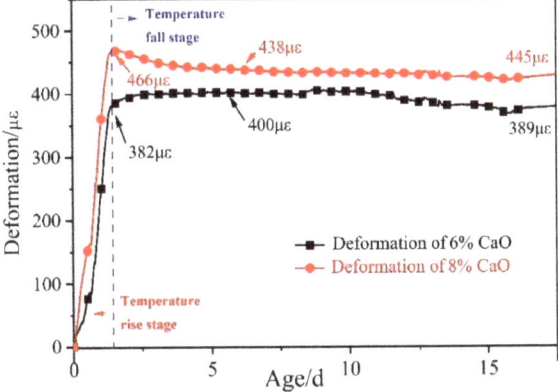

Figure 9. Expansion performance of 6% and 8% CaO expansive agents after deducting deformation of C60-ref.

As further demonstrated by the above experimental results, the expansion effects of 6% and 8% CaO expansive agents were predominantly reflected in the heating stage, while there was no expansion in the cooling stage. The CaO expansive agent might have been consumed by hydration in the early high-temperature heating stage. So, there was almost no expansion in the cooling stage. With the increase in the amount of the CaO expansive agent, the expansion deformation of concrete in the heating stage increased, and consequently, the internal expansion stress of concrete increased. C60 concrete with the 8% CaO expansive agent triggered higher expansion stress in the heating stage, which led to the augmentation of concrete creep in the cooling stage. Accordingly, the expansion amount with the 8% CaO expansive agent appeared to shrink back in the cooling stage.

3.2. Expansion Properties of MgO Expansive Agents in C60 Concrete under Variable Temperature Conditions

As revealed by the above experimental findings, the distinction in the expansion effects of CaO expansive agents at distinct dosing levels was primarily reflected in the heating stage. For C60 concrete structures, CaO expansive agents could only compensate for the early autogenous shrinkage of concrete. The autogenous volume deformation and temperature history of C60 concrete with 0%, 4% 65 s MgO, and 4% 120 s MgO expansive agents subjected to a simulated temperature history are displayed in Figure 10. C60 concrete with the 4% MgO expansive agent was in a state of shrinkage at 24.5 d after the initial setting time. With the initial setting time as the zero point, C60 concrete with the 4% 65 s MgO expansive agent in the heating stage resulted in 422 με expansion deformation. Furthermore, C60 concrete with the 4% 120 s MgO expansive agent gave rise to 359 με expansion deformation. The reaction rate of the highly active MgO (65 s) expansive agent was faster in the heating stage of this temperature history.

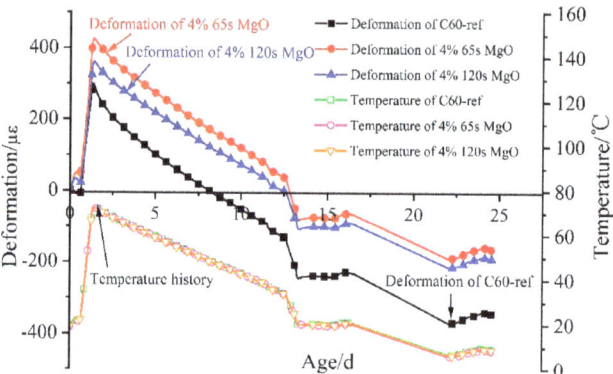

Figure 10. Deformation and temperature history of C60 concrete with 0%, 4% 65 s, and 4% 120 s MgO expansive agents.

The expansion curves of 4% 65 s MgO and 4% 120 s MgO expansive agents in the cooling stage are shown in Figure 11, after deducting the deformation of C60-ref concrete and temperature effects. The deformation of the 4% 120 s MgO expansive agent continued to increase in the cooling stage. The 4% 65 s MgO expansive agent can also produce expansion deformation in the cooling stage. In comparison with the 65 s MgO expansive agent, the 120 s MgO expansive agent triggered continuous expansion and generated 69 με expansion deformation at 15 d in the cooling stage. As the active reaction time rises, the hydration of MgO in the heating process of concrete decreases, and the expansion in the cooling phase increases.

As demonstrated by the above experimental findings, in this temperature history, CaO expansive agents can only compensate for the early concrete shrinkage and store early expansion stress and have no expansion compensation effect in the cooling stage. If CaO expansive agents are employed to increase the amount of concrete expansion, there would be an increment in concrete creep during the cooling stage. Using the single admixtures of MgO expansive agents, C60 concrete with the 4% 65 s MgO or 4% 120 s MgO expansive agent was in a shrinkage state at 24.5 d after the initial setting time and had not reached the expansion or non-shrinkage state. If the experiment raises the amount of MgO to realize the objective of non-shrinkage, the workability of concrete will be seriously decreased, leading to the formation of cavities between the steel pipe and concrete. The experiments of single MgO expansive agents also show that with this temperature history, the activity reaction time of MgO increases, the hydration of MgO in the concrete heating stage decreases, and the expansion of MgO in the cooling stage rises.

3.3. Expansion Properties of Expansive Agents of 6% CaO and 2% MgO with Different Active Times in C60 Concrete under Variable Temperature Conditions

Whether the delayed expansion of MgO could be employed to compensate for the shrinkage of concrete during the cooling stage and realize the compensation for shrinkage deformation in concrete remains for the complete process to be determined. For the sake of selecting suitable CaO and MgO composite expansive agents to compensate for the shrinkage of C60 concrete in the whole process, the expansion deformation laws of expansive agents with 2% MgO (active time of 65 s, 120 s, or 220 s) and 6% CaO composite were compared and analyzed.

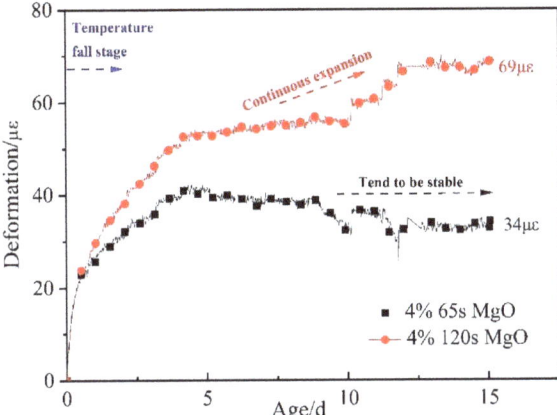

Figure 11. Expansion performance of 4% 65 s MgO and 4% 120 s MgO expansive agents in cooling stage after deducting deformation of C60-ref.

The volume deformation and temperature history of C60 concrete with 0%, 6% CaO, and 6% CaO + 2% MgO expansive agents under simulated variable temperature are depicted in Figure 12. Compared with the base concrete (C60-ref), the expansion deformation of C60 concrete with 6% CaO + 2% MgO increased in the heating stage, while the temperature development history of each proportion of concrete was similar. As a consequence, it could be assumed that the increment in expansion was due to the hydration of expansive agents. With the initial setting time as the zero point, C60 concrete with 6% CaO + 2% 65 s MgO in the heating stage was the highest with 731 με expansion deformation, and the temperature rose to 53.0 °C. Moreover, C60 concrete with the 6% CaO + 2% 120 s MgO expansive agent in the heating stage gave rise to 675 με expansion deformation, and the temperature rose to 48.9 °C. Additionally, C60 concrete with the 6% CaO + 2% 220 s MgO expansive agent in the heating stage triggered 689 με expansion deformation, and the temperature rose to 48.8 °C.

The deformation process of C60 with 0%, 6% CaO, and 6% CaO + 2% MgO expansive agents in the cooling stage is depicted in Figure 13. The shrinkage of C60 concrete with CaO + 2% MgO composite expansive agents was reduced compared to the base concrete (C60-ref) during the cooling stage. The expansive agents of CaO are consumed by hydration in the early heating process, which demonstrates that the expansion in the cooling phase originates from the hydration of MgO, which could undergo expansion during an appropriate cooling course to compensate for the shrinkage. During the cooling stage, deducting the influence of the shrinkage of C60-ref concrete, the 6% CaO + 2% 65 s MgO expansive agent generated 20 με expansion deformation. In addition, the 6% CaO + 2% 120 s MgO expansive agent resulted in 81 με expansion deformation. Most significantly, the 6% CaO + 2% 220 s MgO expansive agent triggered 115 με expansion deformation, indicating that 220 s MgO exhibits the best compensation effect in the cooling stage.

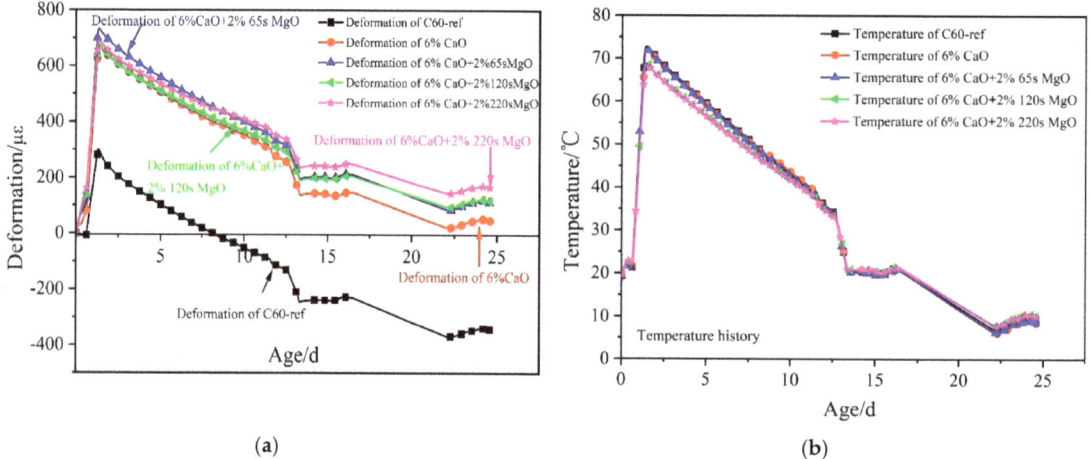

(a) (b)

Figure 12. Deformation and temperature history of C60 concrete with 0%, 6% CaO, and 6% CaO + 2% MgO expansive agents: (**a**) deformation history and (**b**) temperature history.

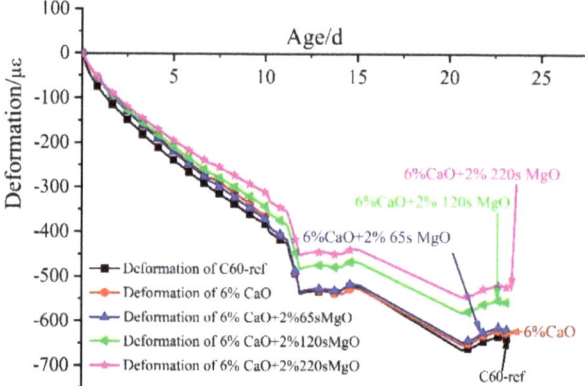

Figure 13. Deformation process of C60 with 0%, 6% CaO, and 6% CaO + 2% MgO expansive agents in cooling stage.

Figure 14 illustrates the expansion deformation of the 6% CaO expansive agent and 6% CaO + 2% MgO expansive agents and the base concrete (C60-ref)'s net deformation and temperature effects, with the initial setting time as the zero point. The expansion effect of CaO expansive agents was predominantly reflected in the heating stage. The expansion deformation during the cooling stage was mainly caused by the MgO expansive agent. Using 120 s MgO and 220 s MgO resulted in continuous expansion during the cooling stage, and the expansion curve did not converge. Using the 220 s MgO expansive agent produced more expansion than 120 s MgO, whereas 65 s MgO did not produce significant expansion deformation in the cooling phase (shown in Figure 14). During the heating process of the concrete, the majority of 2% 65 s MgO may have reacted with water to form brucite, leading to the observed phenomenon. Thus, with the increase in the active reaction time of MgO, the hydration of MgO in the heating stage of concrete decreased, and the expansion of MgO in the cooling stage increased.

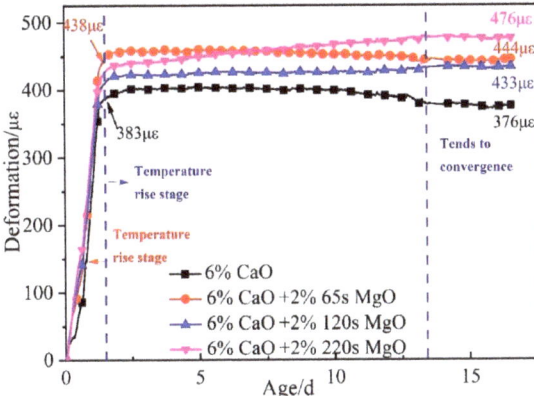

Figure 14. Expansion performance of 6% CaO and 6% CaO + 2% MgO expansive agents after deducting the deformation of C60-ref.

3.4. Expansion Properties of Expansive Agents of 6% CaO and 65 s MgO in Different Proportions in C60 Concrete under Variable Temperature Conditions

The autogenous volume deformation and temperature history of C60 concrete with 0%, 6% CaO + 2% 65 s MgO, and 6% CaO + 4% 65 s MgO expansive agents under simulated variable temperature conditions are shown in Figure 15. With the initial setting time as the zero point, C60 concrete with 6% CaO + 2% 65 s MgO in the heating stage gave rise to 729 με expansion deformation, and the temperature rise was 52.7 °C. Moreover, C60 concrete with 6% CaO + 4% 65 s MgO in the heating stage triggered 953 με expansion deformation, and the temperature rise was 53.0 °C. With the increase in MgO content, the expansion of concrete in the heating stage increases.

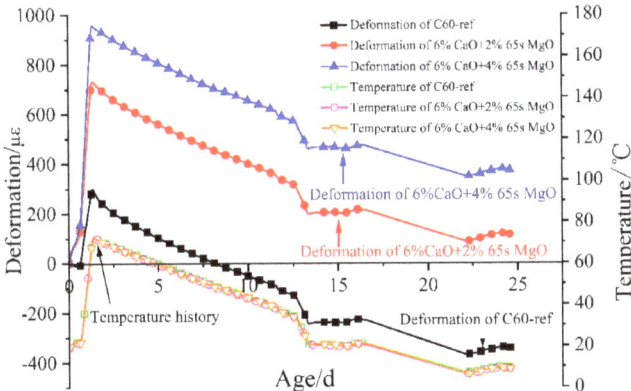

Figure 15. Deformation and temperature history of C60 concrete with 0%, 6% CaO + 2% 65 s MgO, 6% CaO + 4% 65 s MgO expansive agents.

The deformation process of C60 concrete with 0%, 6% CaO + 2% 65 s MgO, and 6% CaO + 4% 65 s MgO expansive agents at in cooling stage is depicted in Figure 16. With the increase in MgO content, the shrinkage of concrete in the cooling stage decreases. The shrinkage of C60 concrete with 6% CaO + 4% 65 s MgO decreased by 35 με at 23 d compared with C60 concrete with 6% CaO + 2% 65 s MgO.

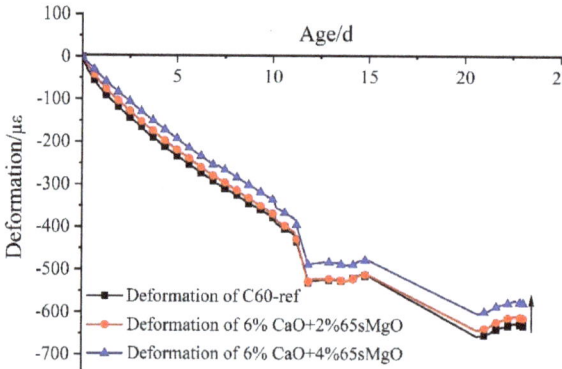

Figure 16. Deformation process of C60 with 6% CaO and 6% CaO + 4% MgO expansive agent at cooling stage.

Figure 17 illustrates the expansion deformation of 6% CaO + 2% 65 s MgO and 6% CaO + 4% 65 s MgO expansive agents and the base concrete (C60-ref)'s net deformation and temperature effects, with the initial setting time as the zero point. The expansion of MgO in the cooling stage increased with the increment in 65 s MgO content. In comparison to 2% 65 s MgO, 4% 65 s MgO continued to expand in the cooling stage of C60 concrete, which demonstrates that expansion can also occur in the cooling stage as the amount of high-activity MgO increases.

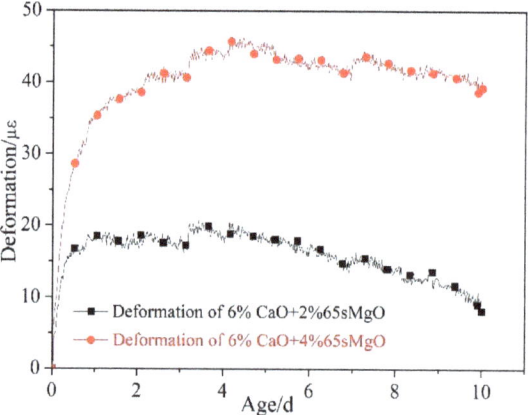

Figure 17. Expansion performance of 6% CaO + 2% MgO and 6% CaO + 4% MgO expansive agents in cooling stage after deducting deformation of C60-ref.

As revealed by the experimental results, the CaO expansive agent has the advantages of a fast hydration rate, large expansion deformation, and a certain amount of expansion stress in concrete in the heating stage. In the cooling stage, concrete-filled steel tubes can achieve a microexpansion or non-shrinkage state by using the delayed expansion property of the MgO expansive agent. The hydration rate of longer-sintered MgO is higher. For the temperature history used, 2% 65 s MgO did not show significant expansion in the cooling stage, while expansion compensation could also be produced in the cooling stage when 65 s MgO expansive agent doping was increased to 4.0%. Under these temperature conditions, 220 s MgO with low activity demonstrated superior compensation for the cooling shrinkage. The expansion performance of 220 s MgO compounded with CaO and

MgO composite expansive agents demonstrated strong temperature sensitivity, which is suitable for compensating for the shrinkage of C60 structural concrete with a high-temperature rise and slow cooling rate.

3.5. Isothermal Calorimetry of Different Expansive Agents

The exothermic courses of hydration of distinct expansive agents in pure water are demonstrated in Figure 18. At a constant temperature of 20 °C, the hydration rates of CaO expansive agents were faster. Furthermore, the accumulated exothermic heat was stabilized within 2 days. The 65 s MgO, 120 s MgO, and 220 s MgO expansive agents exhibited continuous augmentation in the accumulated exotherm over 2 days and yet did not show convergence, and the exothermic rate of hydration was significantly lower than that of CaO expansive agents. Additionally, the exotherm rate of hydration of 65 s MgO was significantly faster than that of 120 s and 220 s MgO. For CaO expansive agents in 20 °C water, the hydration reaction rates are fast, with the basic reaction complete in 2 days. In line with Arrhenius's law, the reaction rate of CaO and MgO expansive agents must increase exponentially when the temperature rises. It is deduced that the CaO expansive agents completely reacted in the heating stage of this temperature history of C60 core concrete in the steel tube arch.

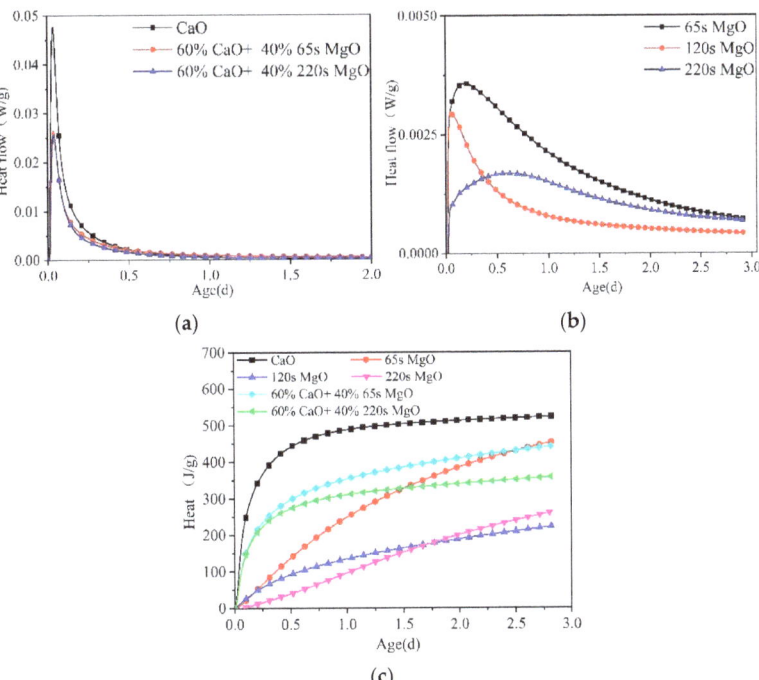

Figure 18. Hydration heat liberation of distinct expansive agents in pure water (paste) at 20 °C: (**a**) heat flow of CaO and CaO + MgO, (**b**) heat flow of MgO, and (**c**) total heat release of the expansive agent.

3.6. Hydration Degree of MgO Expansive Agent in Cement Paste under Variable Temperature Conditions

After being cured under variable temperature conditions, as shown in Figure 7, the XRD patterns of the cement paste (4% MgO, 16% fly ash, 80% cement, and water–binder ratio of 0.29) mixed with the 4% 120 s MgO expansive agent and the cement sample without an expansive agent (20% fly ash, 80% cement, and water–binder ratio of 0.29) were obtained and are shown in Figure 19. It can be seen that at 24.5 d after the initial setting time, the cement paste with the 4% 120 s MgO expansive agent contained MgO and Mg(OH)$_2$ minerals. Table 3 shows the hydration degree of MgO in cement pastes with 6% CaO + 4% 65 s MgO, 4% 65 s MgO, and 4% 120 s MgO expansive agents at variable temperatures. At the same age, the hydration degree of 65 s MgO was higher than that of the 120 s MgO expansive agent. This shows that highly active MgO has higher hydration activity and a higher hydration rate. Whether the CaO expansive agent was added to the cement paste had little effect on the hydration degree of MgO. Under this variable temperature condition, more than 85% of the 65 s MgO expansive agent was almost completely hydrated. About 30% of MgO in the cement paste mixed with the 4% 120 s MgO expansive agent was not hydrated. The hydration degree of the MgO expansive agent is consistent with the change rule of the expansion amount of the MgO expansive agent (shown in Figure 10). The MgO expansive agent with high activity (65 s MgO) undergoes a large expansion in the early stage and a small expansion in the later stage. The MgO expansive agents with low activity (120 s MgO and 220 s MgO) undergo a small expansion in the early stage and a large expansion in the later stage.

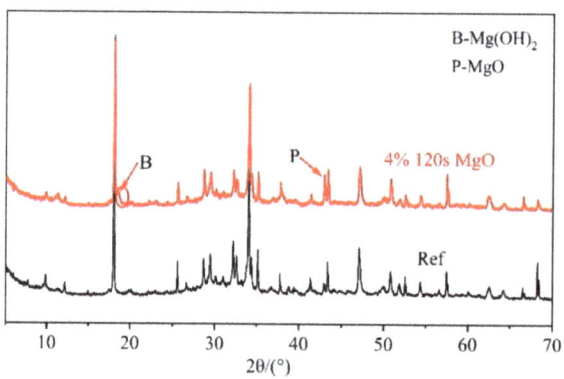

Figure 19. Comparison of XRD patterns of cement pastes with and without MgO expansive agent.

Table 3. Hydration degree of MgO in cement paste under variable temperature conditions.

No.	6% CaO + 4% 65 s MgO	4% 65 s MgO	4% 120 s MgO
Hydration degree/%	87.3	85.5	70.3

3.7. SEM Analyses of Cement Paste Mixed with Expansion Agent

After being cured under variable temperature conditions, as shown in Figure 7, the morphologies of cement pastes with 0%, 4%, and 6% CaO expansive agents were observed by SEM and are shown in Figure 20. A large number of hexagonal plate Ca(OH)$_2$ crystals appeared in the cement paste after adding the CaO expansive agent. As the content of the CaO expansive agent increased, the content of Ca(OH)$_2$ generated by hydration also increased. As can be seen in Figure 20c, microcracks appeared in the cement paste with the 6% CaO expansive agent. The formation of microcracks may be caused by the volume expansion of Ca(OH)$_2$ generated by the hydration of the CaO expansive agent. Microcracks were not observed in the cement paste containing the 4% CaO expansive agent. The reason for this could be that the crystallization pressure of Ca(OH)$_2$ crystals generated by the 4% CaO expansive agent in the cement paste does not exceed the tensile strength of the cement paste [43].

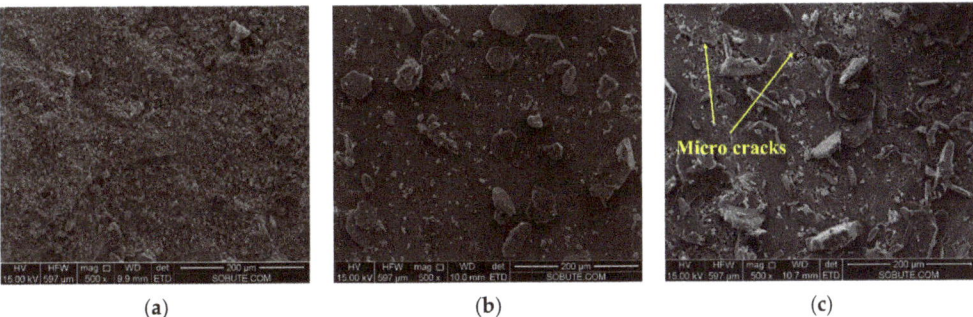

Figure 20. Morphologies of cement pastes with 0%, 4%, and 6% CaO expansive agents: (**a**) 0% CaO expansive agent, (**b**) 4% CaO expansive agent, and (**c**) 6% CaO expansive agent.

Figure 21 shows Ca(OH)$_2$ with a hexagonal plate structure formed by the hydration of the CaO expansive agent. The energy spectrum test results of EDS Spot 2 in the microscopic picture of the cement paste with the 6% CaO expansive agent show that the hexagonal plate material consists of Ca(OH)$_2$ crystals [44] (shown in Figure 22).

Figure 21. Ca(OH)$_2$ with hexagonal plate structure formed by hydration of CaO expansive agent (×10,000).

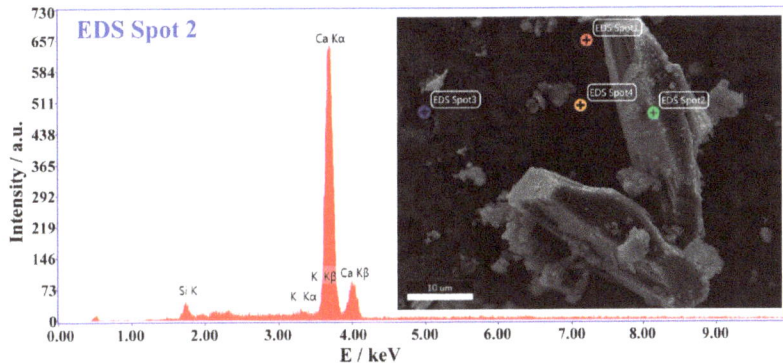

Figure 22. The energy spectrum test results of EDS Spot 2 in the microscopic picture of cement paste with 6% CaO expansive agent.

After being cured under variable temperature conditions, as shown in Figure 7, the morphologies of cement pastes with 4% MgO expansive agents were observed and are shown in Figure 23. No obvious microcracks were found in the cement paste with the 4% MgO expansive agent. MgO and hydrated Mg(OH)$_2$ are not easy to observe under an electron microscope. The energy spectrum test results of EDS Spot 5 in the microscopic picture of the cement paste with the 4% 120 s MgO expansive agent are shown in Figure 24. At EDS Spot 5, Mg(OH)$_2$ was formed.

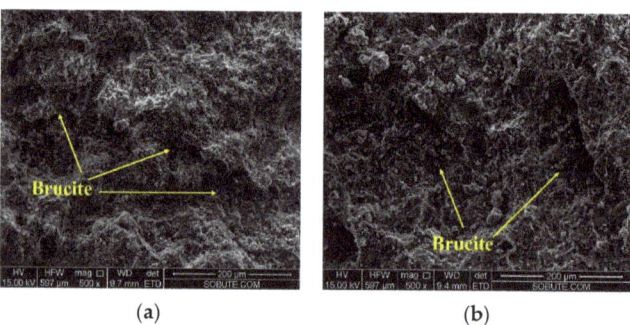

Figure 23. Morphologies of cement pastes with 4% MgO expansive agents: (**a**) 4% 65 s MgO and (**b**) 4% 120 s MgO.

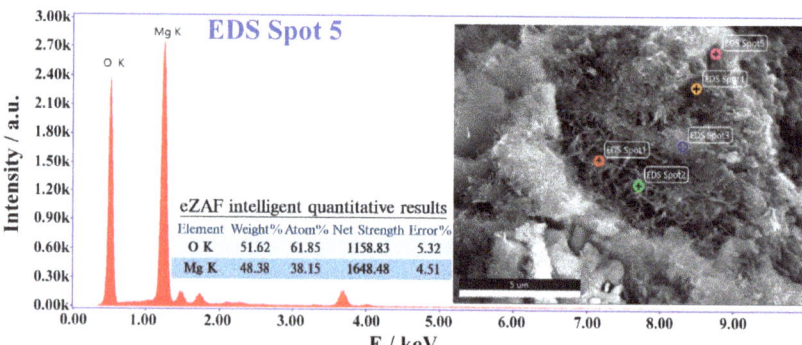

Figure 24. The energy spectrum test results of EDS Spot 5 in the microscopic picture of cement paste with 4% 120 s MgO expansive agent.

After being cured under variable temperature conditions, as shown in Figure 7, the morphology of the cement paste with the 6% CaO + 2% 220 s MgO expansive agent was observed and is shown in Figure 25. As can be seen from the figure, a large number of hexagonal plate Ca(OH)$_2$ crystals appeared in the cement paste, and microcracks appeared in the cement paste with the 6% CaO and 2% 220 s MgO expansive agent.

The microscopic analysis shows that the CaO expansive agent can produce a large number of hexagonal plate Ca(OH)$_2$ crystals in the cement paste. Microcracks appeared in the cement paste with the 6% CaO expansive agent. The formation of microcracks may be caused by the volume expansion of Ca(OH)$_2$ generated by the hydration of the CaO expansive agent. In the cement paste with the 4% MgO expansive agent, no obvious microcracks were found, whereas microcracks appeared in the cement paste with the 6% CaO and 2% 220 s MgO expansive agent.

3.8. Discussion and Analysis

As mentioned above, different types of expansive agents have different hydration and expansion characteristics. In actual projects, the autogenous shrinkage and temperature-fall-induced shrinkage generated in C60 concrete in the early and middle hydration stages are large, with a wide temperature variation range. The large expansion of concrete should be compensated. In addition, the later autogenous shrinkage and temperature-fall-induced shrinkage of concrete are smaller, and microexpansion is needed to compensate for them, thus inhibiting their shrinkage to stabilize the expansion precompression stress formed in the early stage. Finally, shrinkage-free concrete should be achieved in all hydration stages.

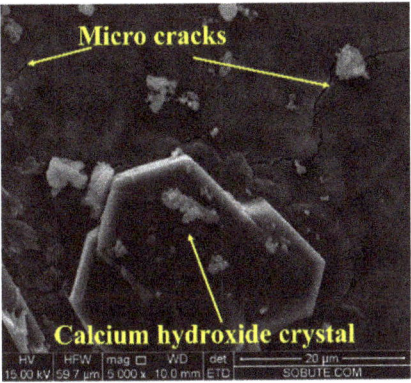

Figure 25. Morphology of cement paste with 6% CaO + 2% 220 s MgO expansive agent (×5000).

The CaO expansive agent is beneficial to early expansion, while the MgO expansive agent can achieve middle- and late-age expansion; thus, a potentially good method to realize shrinkage-free concrete throughout the whole process is to design multicomponent expansive agents with CaO and MgO compounds.

In this study, the influence of the temperature history of core concrete in a steel tube arch on the expansion of C60 concrete with specific-activity MgO and CaO expansion components in certain proportions, a single CaO expansive agent, and a single MgO expansive agent in C60 concrete was investigated. The experimental results showed that the expansion effect of CaO expansive agents was predominantly reflected in the heating stage, while there was no expansion in the cooling stage. The reason for this is that CaO expansive agents completely reacted in the heating stage of this temperature history of C60 core concrete in a steel tube arch, according to microthermal experimental results. Moreover, the microscopic morphology showed that the expansion energy of the CaO expansive agent is larger. Microcracks appeared in the cement paste with the 6% CaO expansive agent. Liu et al. [45] found that the hydration degree of a CaO expansive agent is large at early ages, such that the hydration degree is 39.94% after being hydrated for 0.5 h at 10 °C in pure water, while the hydration degree is 52.47% after being hydrated for 5 min at 40 °C. Xia et al. [46] also stated that the hydration rate of the CaO expansive agent is fast in the early stages. The expansion deformation in the cooling stage was mainly caused by the MgO expansive agent. The 120 s MgO and 220 s MgO expansion agents resulted in continuous expansion during the cooling stage, and the expansion curve did not converge. The 220 s MgO expansive agent produced more expansion than 120 s MgO. The 65 s MgO expansive agent did not produce significant expansion deformation in the cooling phase. During the heating process of the concrete, the majority of 2% 65 s MgO reacted with water to form brucite, leading to the observed phenomenon. With the increase in the active reaction time of MgO, the hydration of MgO in the heating stage of concrete decreased, and the expansion of MgO in the cooling stage increased.

Following the actual temperature history of concrete, selecting the proper activity of MgO expansive agents and the ratio of CaO and MgO in multicomponent expansive agents could compensate for concrete shrinkage throughout the whole process. Zhao et al. [47] also found that a CaO and MgO composite expansive agent can compensate for concrete shrinkage not only during the early stage but also during the later stage at normal temperatures. Thus, the CaO and 220 s MgO composite expansive agent shows strong temperature sensitivity, which is suitable for compensating for the shrinkage of concrete in the case of a high-temperature rise and slow cooling rate. Furthermore, the CaO and 120 s MgO composite expansive agent is suitable for compensating for the shrinkage of concrete in the case of a low-temperature rise and fast cooling rate.

4. Conclusions

Utilizing appropriate expansive agents during cement hydration is one of the main techniques to compensate for concrete shrinkage and prevent voids and debonding between the steel pipe and core concrete in concrete-filled steel tubes. In this work, calcium oxide and magnesium oxide composite expansive agents were designed with different calcium/magnesium ratios and various types of MgO activities with different calcination temperatures and times. Their expansion and hydration properties in C60 concrete under variable temperature conditions were investigated, with the aim of simulating the real construction process. Subsequently, the effect of the calcium–magnesium ratio and magnesium oxide activity on deformation was analyzed. The main conclusions are summarized as follows:

(1) The expansion effects of 6% and 8% CaO expansive agents were predominantly reflected in the heating stage (from 20.0 °C to 72.0 °C at 3 °C/h), while there was no expansion in the cooling stage (from 72.0 °C to 30.0 °C at 3 °C/d, and then to 20.0 °C at 0.7 °C/h). The CaO expansive agent was hydrated in the early high-temperature heating stage, resulting in almost no expansion in the cooling stage. With the increase in the amount of the CaO expansive agent, the expansion deformation of concrete in the heating stage increased, and consequently, the internal expansion stress of concrete increased.

(2) The expansion deformation in the cooling stage was mainly caused by the MgO expansive agent. The 120 s MgO and 220 s MgO expansive agents resulted in continuous expansion during the cooling stage, and the expansion curve did not converge. The 220 s MgO expansive agent produced more expansion than 120 s MgO. During the heating process of the concrete, the majority of 2% 65 s MgO reacted with water to form brucite in large amounts, leading to its lower expansion deformation in the later cooling process. With the increase in the active reaction time of MgO, the hydration of MgO in the heating stage of concrete decreased, and the expansion of MgO in the cooling stage increased.

(3) The construction and building temperatures show a remarkable influence on the expansion performance of CaO and MgO composite expansive agents. In actual projects, the temperature history of concrete varies immensely from structure to structure on account of many factors, such as the material, environment, and structural dimensions. In accordance with the actual temperature history of concrete, selecting the proper activity of MgO and the ratio of CaO to MgO could compensate for concrete shrinkage throughout the whole process. The CaO and 220 s MgO composite expansive agent shows strong temperature sensitivity, which is suitable for compensating for the shrinkage of concrete in the case of a fast high-temperature rise and a slow cooling rate.

This work will guide the application of different types of CaO-MgO composite expansive agents in concrete-filled steel tube structures. More work should be conducted in the future to build a theoretical model of temperature effects on different components of the expansive agent to guide engineering projects.

Author Contributions: A.L.: conceptualization, methodology, investigation, software, project administration, and writing—original draft; W.X.: data curation, formal analysis, investigation, conceptualization, resources, and supervision; Q.W.: data curation, formal analysis, supervision, and writing—review and editing; R.W.: supervision and project administration; Z.Y.: investigation, validation, and writing—review and editing. All authors have read and agreed to the published version of the manuscript.

Funding: Science and Technology Research and Development Program of China Railway Corporation (2017G006-B); Natural Science Foundation of Jiangsu Province, China (BK20221198).

Institutional Review Board Statement: Not applicable.

Informed Consent Statement: Not applicable.

Data Availability Statement: The datasets generated during and/or analyzed during the current study are available from the corresponding author upon reasonable request. All data generated or analyzed in this research are included in this published article. Additionally, readers can access all data used to support the conclusions of the current study from the corresponding author upon request.

Conflicts of Interest: The authors declare no conflict of interest.

References

1. Chen, B.; Liu, J. Review of construction and technology development of arch bridges in the world. *J. Traffic Transp. Eng.* **2022**, *1*, 27–41. (In Chinese)
2. Younas, S.; Hamed, E.; Uy, B. Effect of creep on the strength of high strength concrete-filled steel tubes. *J. Constr. Steel Res.* **2023**, *201*, 107719. [CrossRef]
3. Han, X.; Han, B.; Xie, H.; Yan, W.; Yu, J.; He, Y.; Yan, L. Seismic stability analysis of the large-span concrete-filled steel tube arch bridge considering the long-term effects. *Eng. Struct.* **2022**, *268*, 114744. [CrossRef]
4. Patel, V.I. Analysis of uniaxially loaded short round-ended concrete-filled steel tubular beam-columns. *Eng. Struct.* **2020**, *205*, 110098. [CrossRef]
5. Zheng, J.; Wang, J. Concrete-Filled Steel Tube Arch Bridges in China. *Engineering* **2018**, *4*, 143–155. [CrossRef]
6. Yang, K.; Gao, L.; Zheng, K.; Shi, J. Mechanical behavior of a novel steel-concrete joint for long-span arch bridges—Application to Yachi River Bridge. *Eng. Struct.* **2022**, *265*, 114492. [CrossRef]
7. Chen, B.; Wang, T. Overview of concrete filled steel tube arch bridges in China. *Pract. Period. Struct. Des. Constr.* **2019**, *14*, 70–80. [CrossRef]
8. Li, Y.; Wang, Z.; Li, D. Mechanical behavior of concrete-filled steel tubular columns with initial concrete imperfection under long-term sustained load. *J. Build. Struct.* **2020**, *10*, 112–120. (In Chinese)
9. Liao, F.; Han, H.; Wang, Y. Cyclic behaviour of concrete-filled steel tubular (CFST) members with circumferential gap under combined compression-bending-torsion load. *China Civ. Eng. J.* **2019**, *7*, 57–80.
10. Wang, Z.; Han, J.; Wei, J.; Lu, J.; Li, J. The axial compression mechanical properties and factors influencing spiral-ribbed thin-walled square concrete-filled steel tube composite members. *Case Stud. Constr. Mater.* **2022**, *17*, e01510. [CrossRef]
11. Peng, Y.; Qiang, S.; Liu, Y. Study of sunshine temperature distribution in circular concrete-filled steel tube arch rib. *Bridge Constr.* **2006**, *6*, 18–20. (In Chinese)
12. Zhou, X.; Zhan, Y.; Mou, T.; Li, Z. Experimental Research on Flexural Mechanical Properties of Ultrahigh Strength Concrete Filled Steel Tubes. *Materials* **2022**, *15*, 5262. [CrossRef] [PubMed]
13. Su, J. Survey on CFST Arch Bridges and Research on Their Void Problem. Master's Thesis, Southwest Jiaotong University, Chengdu, China, 2012. (In Chinese)
14. Han, L.; Yang, Y.; Li, Y.; Feng, B. Hydration heat and shrinkage of high performance concrete-filled steel tubes. *China Civ. Eng. J.* **2006**, *3*, 1–9. (In Chinese)
15. Chang, X.; Huang, C.; Jiang, D.; Song, Y. Push-out test of pre-stressing concrete filled circular steel tube columns by means of expansive cement. *Constr. Build. Mater.* **2009**, *23*, 491–497.
16. Nagataki, S.; Gomi, H. Expansive admixtures (mainly ettringite). *Cem. Concr. Compos.* **1998**, *20*, 163–170. [CrossRef]
17. Carballosa, P.; García Calvo, J.L.; Revuelta, D. Influence of expansive calcium sulfoaluminate agent dosage on properties and microstructure of expansive self-compacting concretes. *Cem. Concr. Compos.* **2020**, *107*, 103464. [CrossRef]
18. Yu, Z.; Zhao, Y.; Ba, H.; Liu, M. Synergistic effects of ettringite-based expansive agent and polypropylene fiber on early-age anti-shrinkage and anti-cracking properties of mortars. *J. Build. Eng.* **2021**, *39*, 102275. [CrossRef]
19. Deng, M.; Hong, D.; Lan, X.; Tang, M. Mechanism of expansion in hardened cement pastes with hard-burnt free lime. *Cem. Concr. Res.* **1995**, *2*, 440–448. [CrossRef]
20. Yang, G.; Wang, H.; Wan-Wendner, R.; Hu, Z.; Liu, J. Cracking behavior of ultra-high strength mortar with CaO-based expansive agent and superabsorbent polymer. *Constr. Build. Mater.* **2022**, *357*, 129281. [CrossRef]
21. Zhang, J. Recent advance of MgO expansive agent in cement and concrete. *J. Build. Eng.* **2022**, *45*, 103633. [CrossRef]
22. Jiang, F.; Mao, Z.; Yu, L. Hydration and expansion characteristics of MgO expansive agent in mass concrete. *Materials* **2022**, *15*, 8028. [CrossRef]
23. Mo, L.; Deng, M.; Tang, M.; Al-Tabbaa, A. MgO expansive cement and concrete in China: Past, present and future. *Cem. Concr. Res.* **2014**, *57*, 1–12. [CrossRef]
24. Zhou, Q.; Lachowski, E.E.; Glasser, F.P. Metaettringite, a decomposition product of ettringite. *Cem. Concr. Res.* **2004**, *34*, 703–710. [CrossRef]
25. Baquerizo, L.; Matschei, T.; Scrivener, K. Impact of water activity on the stability of ettringite. *Cem. Concr. Res.* **2016**, *79*, 31–44. [CrossRef]

26. Shen, P.; Lu, J.; Zheng, H.; Lu, L.; Wang, F.; He, Y. Expansive ultra-high performance concrete for concrete-filled steel tube applications. *Cem. Concr. Compos.* **2020**, *114*, 103813. [CrossRef]
27. Wang, Z.; Ding, J.; Cai, Y.; Ning, F. Research progress on surface modification of calcium hydroxide expansive additive. *Bull. Ceram. Soc.* **2017**, *1*, 121–125. (In Chinese)
28. Feng, J.; Miao, M.; Yan, P. The effect of curing temperature on the properties of shrinkage-compensated binder. *Sci. China Technol. Sci.* **2011**, *54*, 869–875. (In Chinese) [CrossRef]
29. Wang, N.; Xiu, X. Study on the application of calcium-oxide calcium sulphoaluminate composite expansion agent in high performance concrete. *Constr. Technol.* **2017**, *S1*, 225–257. (In Chinese)
30. Mo, L.; Deng, M.; Tang, M. Effects of calcinations condition on expansion property of MgO-type expansive agent used in cement-based materials. *Cem. Concr. Res.* **2010**, *40*, 437–446. [CrossRef]
31. Li, S.; Cheng, S.; Mo, L.; Deng, M. Effects of Steel Slag Powder and Expansive Agent on the Properties of Ultra-High Performance Concrete (UHPC): Based on a Case Study. *Materials* **2020**, *13*, 683. [CrossRef]
32. Xu, K.; Liu, P.; Min, Q.; Yang, J. Concrete structural self-waterproofing system of underground works and its engineering application. *China Build. Waterproofing* **2020**, *11*, 37–41.
33. Chen, C.; Lin, X. Magnesium Oxide Expansion Agent and Its Application in Concrete. *Sci. Technol. Eng.* **2020**, *28*, 11413–11420. (In Chinese)
34. Yao, H.; Zheng, J.; Xue, X.; Li, H. Experimental research on using MgO expansion agent self-stressing concrete-filled steel tube. *J. Jiamusi Univ. (Nat. Sci. Ed.)* **2008**, *3*, 289–291. (In Chinese)
35. Cai, Y. Research on the Preparation and Performance of Magnesia Micro-Expansion C50 Steel Tube-Confined Concrete. Master's Thesis, Wuhan University of Technology, Wuhan, China, 2008. (In Chinese)
36. Liu, J.; Zhang, S.; Tian, Q.; Guo, F.; Wang, Y. Deformation of high performance concrete containing MgO composite expansive agent. *J. Southeast Univ. (Nat. Sci. Ed.)* **2010**, *40*, 150–154. (In Chinese)
37. Yu, F.; Feng, J.; Wang, S.; Yang, G.; Yan, P. Study on expansion and mechanical properties of composite cementitious systems with multi-expansion sources expansion agent. *Bull. Ceram. Soc.* **2019**, *1*, 148–154. (In Chinese)
38. Zhao, H.; Li, X.; Chen, X.; Qiao, C.; Xu, W.; Wang, X.; Song, H. Microstructure evolution of cement mortar containing MgO-CaO blended expansive agent and temperature rising inhibitor under multiple curing temperatures. *Constr. Build. Mater.* **2021**, *278*, 122376. [CrossRef]
39. GB175-2007; Standard for Ordinary Portland Cement. China Architecture & Building Press: Beijing, China, 2007. (In Chinese)
40. DL/T 5296-2013; Technical Specification of Magnesium Oxide Expansive for Use in Hydraulic Concrete. China Electric Power Press: Beijing, China, 2020. (In Chinese)
41. Lu, A.; Xu, W.; Wang, R.; Wang, Y.; Tian, Q.; Liu, J. Interpretation of T/CECS 10082—2020 Calcium and Magnesium Oxides Based Expansive Agent for Concrete. *China Concr. Cem. Prod.* **2020**, *9*, 74–78. (In Chinese)
42. Miao, C.; Tian, Q.; Sun, W.; Liu, J. Water consumption of the early-age paste and the determination of "time-zero" of self-desiccation shrinkage. *Cem. Concr. Res.* **2007**, *37*, 1496–1501.
43. Mu, S.; Sun, Z.; Sun, X. A Study on the Microstructure and Expanding Mechanism of Highly Free-calcium Oxide Cement. *J. Wuhan Univ. Technol.* **2001**, *23*, 27–29. (In Chinese)
44. Chen, X.; Wei, S.; Wang, Q.; Tang, M.; Shen, X.; Zou, X.; Shen, Y.; Ma, B. Morphology prediction of portlandite: Atomistic simulations and experimental research. *Appl. Surf. Sci.* **2020**, *502*, 144296. [CrossRef]
45. Liu, J.; Guo, S.; Tian, Q.; Wang, Y.; Zhang, S. Hydration of CaO Expansion Clinker. *J. Build. Mater.* **2014**, *17*, 15–18. (In Chinese) [CrossRef]
46. Xia, R.; Wang, H.; Xiang, F.; Wang, H.; Zhang, Z.; Cheng, F. Influencing Factors of Carbonation Modification of Calcium Oxide Expansive Clinker. *Mater. Rep.* **2022**, *36*, 22080160. (In Chinese)
47. Zhao, H.; Xiang, Y.; Chen, X.; Huang, J.; Xu, W.; Li, H.; Wang, Y.; Wang, P. Mechanical properties and volumetric deformation of early-age concrete containing CaO-MgO blended expansive agent and temperature rising inhibitor. *Constr. Build. Mater.* **2021**, *299*, 123977. [CrossRef]

Disclaimer/Publisher's Note: The statements, opinions and data contained in all publications are solely those of the individual author(s) and contributor(s) and not of MDPI and/or the editor(s). MDPI and/or the editor(s) disclaim responsibility for any injury to people or property resulting from any ideas, methods, instructions or products referred to in the content.

MDPI
St. Alban-Anlage 66
4052 Basel
Switzerland
www.mdpi.com

Materials Editorial Office
E-mail: materials@mdpi.com
www.mdpi.com/journal/materials

Disclaimer/Publisher's Note: The statements, opinions and data contained in all publications are solely those of the individual author(s) and contributor(s) and not of MDPI and/or the editor(s). MDPI and/or the editor(s) disclaim responsibility for any injury to people or property resulting from any ideas, methods, instructions or products referred to in the content.

www.ingramcontent.com/pod-product-compliance
Lightning Source LLC
LaVergne TN
LVHW070657100526
838202LV00013B/987